Unsung Heroes

LEE MADDOX
'1952'

California Vaquero

Arnold R. Rojas

CALIFORNIA VAQUERO

BY

A.R. Rojas

This book is a complete, unabridged edition of the original classic by Arnold R. Rojas (1896-1988). We have kept it exactly as the edition he published in 1953. Rojas was 57.

The publisher wishes to thank Mr. Rojas' family, especially Ms. Juanita Montes for help in this project.
We thank you, the reader, for your interest in the work of this important author.

William C. Reynolds, Publisher
Alamar Media, Inc.
Santa Ynez, California
www.oldcowdogs.com

Printed in the United States of America

ISBN: 978-0-9890701-2-6

Foreward . . .

We have judged it unwise to rewrite this narrative or even to introduce more order into it. The story flows just as it would around a campfire after a busy day. Who are we to dare suggest that the Last Vaquero submit to literary rules. We want to live with the remnant of the vaqueros and to see them just as they are. Let us then have the good sense to permit one of them to talk just as he wills. Let him remain a vaquero so that we may know him as he is. He will keep us interested just as he was kept interested as a kid by the old Mexican and Californian worshipers of the horse, the cattle, and the lesser denizens of the California plains and hills at the end of the 19th and the beginning of the 20th Century. James H. Culleton, Editor

My Thanks . . .

My sincere gratitude goes to all the many and good people who helped make this book possible. High on a long list, which must remain incomplete, are the following: Walter Kane, Bernice Harrell Chipman, and Ralph Krieser of the Bakersfield *Californian*; Frank Latta and Richard Bailey of the Kern County Historical Society; Elmer Belloumini and Louise Renfro. Then, too, we have the old vaqueros: Leonardo Chavarria, Emliano Castro, Orville Glen, Guadalupe Gomez, Adolfo Encinas, and many others who with a kind word encouraged me. Jim Day, of course, comes in for very special *gracias*. Was it not he who bore with my defects and opened up for me the gate to a new world — the enjoyment of the fruits of the pen!

Dedication . . .

This book is affectionally dedicated to those *Don Quijotes* who in yesteryear rode the ranges of Kern County carrying a long rope instead of a lance and wearing a *ten gallon hat* in place of the *Mambrinos* helmet. They upheld the vaquero tradition with the tenacity, he of the *Sorrowful Figure,* that of the knight errant, looking back over the bottomless canon of the years I see much of Don Quijote and naught of Sancho Panza in the range rider of long ago.

Arnold R. Rojas
Bakersfield, California
September 4, 1952

Adolfo Encinas

Introduction . . .

This is a collection of tales gathered from the vaqueros of the San Joaquin Valley and its surrounding Sierra. The time: The years when the greatest cattle empires in North America were passing out of existence.

From four old men I gained a passion for the stories old men tell. They taught me that there is much pleasure and knowledge to be found in paying heed to the words of those who have gleaned wisdom from experience. In after years in the bunkhouse or lonely line shack, at the campfire or riding the range stirrup to stirrup, I was to find the vaquero quite often a peerless story teller. I learned too that there can be much food for study and a wealth of interesting conjecture in an old man sitting in the sun, a saddle marked horse dozing in the corner of a corral, an old bridle, a rusty spur, or an adobe wall crumbling into dust.

Later I was to find on the cattle ranges many other vaqueros who knew, in more or less degree, something of this folklore which has been left to die of neglect in California.

They told tales which were basically the same and like the vaquero's bit, spur, and much of his language, had their origin in the tents of desert horsemen.

The Story Begins . . .

The vaquero with his great rowelled spurs, rawhide reata, Spanish bit and tireless mustang is gone from the great valley into the limbo of forgotten things. He will never return, for the sand-lapper has at last won the land. The cotton and potato fields like tentacles have crept on and over the flat grass lands to the foot of the sierra which stands like a barrier and frowns down on this mutation as if in fear of a like fate.

There, at the end of Comanche Point which looms like an island in a sea of cotton bolls, the scrawny old fig tree, stunted from arid seasons, lone relic of an old stage station in another era, and later a line camp of Tejon, stands in the wet rows of a green field having found that rare thing, abundance in old age.

The gate below San Emidio, where one spring morning 9000 head of steers were counted through on the drive to summer pasture, has long since fallen. Some migrant used its timbers to warm his shivering body. The southern San Joaquin is no longer a cattle country. Of the great army of old school vaqueros who rode the Kern ranges, all that are left are three aged, decrepit men living their few remaining years in Bakersfield: the end of an era that had so courageous a beginning.

Far to the east lies Tubac sleeping in the sun, its houses sprawl amid the mounds of earth, once walls of its proud presidio, careless and indifferent of its past glory when the standard of Aragon and Castile flew over this far outpost. The tide of Iberian Empire rolling north halted here before the wall of Indian menace then turned west, built this tiny village to crouch among the red clay hills, like a dove from a circling hawk.

The desolate California desert, so inhospitable to the early wayfarer, was feared less by the iron men of Spain than was the most consummate fighting man in the Western Hemisphere, the Apache warrior. Though his character has been defamed, and he has been lied about with evil malice and rancor, if there were a special hall of fame for indomitable stalwarts who fight for lost causes and lead forlorn hopes, the Apache deserve a place with other tribesmen; Sioux, Modoc, Yaqui, Comanche and his kinsman, the Aztec, who stood firm and gave valiant battle to Iberian and Anglo-Saxon alike in defense of their homeland.

To the east and west great cities flourish, founded on the industry that had its inception in this little pueblo of Tubac.

There is a spirit of dogged tenacity which runs like a vein through the story, and characterizes the laborious founding of the cattle industry in California, and subsequently from here, the entire western United States.

The story begins with Father Kino, founder of cattle ranches among Papagos and Pimas in defiance of Apaches, and with DeAnza, one of the early day Spaniards who ranked in ability and astuteness with Cortez, and possibly excelled him in some respects. DeAnza's policy in dealing with the aborigines is responsible for the survival of the southwestern Indians.

A grandee of Spain, he became California's first trail drive boss when he left the lonely frontier out-post of Tubac to cross the barrier of desert sand dunes with cattle for the missions over the *camino del diablo* through a land truly God forsaken and probably designed by the Creator to teach men, compelled to venture over it, the meaning of hell.

Perhaps no system other than that of the padres could have propagated Spanish cattle and horses so thoroughly over the seven western states as that which began here with the handful brought into this state in that first drive. This was the first overland East-West drive. Also the first drive of settlers' cattle. Much mission stock had come from Vellicata, Lower California.

Later as the few were multiplying into countless thousands on ranges that were to teem with cattle, following in the trail-blazers' footsteps came men with their women and children, whose descendants were to be the vaqueros of California, and the drivers of trail herds into Nevada, Utah and Wyoming, north into Oregon, to stock the ranges of those states. They plodded over the *"Road of the Devil"* riding what they could get; horses sometimes an ox from the cholla-eating bands of cattle that range the desert west of the Sonorita river in northern Sonora, but most often their mounts were burros.

The Burro . . .

Though their masters may never have thought to eulogize them, the real heroes of the migrations into California from Sonora were the lowly burros, half dead bags of bones, specters of famine and over-work, with feet worn down to the quick at the end of the journey. Masters and beasts were examples of the ultimate in human and animal misery after traveling a road, one stretch of which was 130 miles without running water, 60 with no water of any kind in all its burning length.

It is difficult to see how the people who came in early days could have made the journey in such numbers without the patient little animal the paisano could most readily afford.

There is a numerous and much respected family in Kern county whose forebears, a father and seven sons, walked while the mother rode an ox, the only animal they possessed.

Could he have known, many a far-venturing prospector, in later years gambling his life on the stamina of his burro in a thirsty land, would have been thankful that the ancestors of his sturdy little beast has been tested in the crucible of the camino, and had not been found wanting. Many rich mines in the deserts of the west would never have been discovered had not the early-day prospector found at hand a noble little animal which fools, in the depth of their ignorance, have made the symbol of stupidity.

One of the desert rats' most popular tales is the yarn of the strayed burro which was found by its owner standing over a rich deposit of gold-bearing ore—tales which no doubt resulted from the efforts of some prospector to impress scoffers with the worth and sagacity of his partner and companion.

The rancheros of the old days, when the time came to slaughter cattle for their hides, would hold a *rodeo de escoger y desechar* (round-up to select and discard). Old *paisanos* say that when the final great rodeo comes the burrito will be the first to be parted out into the *parada* of the *escojidos* (selected ones).

Ask any old *paisano* who knows him and he will tell you the burro is the smartest and bravest of all animals. Pit him against the largest and strongest horse and the proudest charger will turn and flee from the onslaught of the lowly jackass. He will, in a fight with a grizzly bear, give a better account of himself than any other animal, battling with a ferocity that belies his melancholy bray, gentle nature, and great trusting eyes. He never trusts man altogether, but relies on himself when in danger. His instinct tells him when man is in error.

One may put a loop around his neck and between his legs, so he will not choke, and tie a wild stallion, bull or steer to him with five feet or so of rope between him and the captive, turn them loose and it will be the wild one who will submit, never the burro, for when the captive runs, the burro will run, if it heads for the brush, he will stop and hold it until it heads in the direction the burro wishes to go. If the wild one tries to paw, kick, or hook him the jackass will duck, dodge, or run, but never get hurt, and pull his charge along, and in the end deliver a much subdued horse, bull, or steer at the home corral.

Mule and Burro . . .

The mule who has a lordly contempt for man's reasoning powers, and obeys him only so long as man does not put him in danger, inherits his brains from the burro's side of his family. For unlike the horse, he will not allow himself to be ridden into carnage in war nor will he like the horse, stand and obey the rein and be killed by his master's wish, but uses his instinct to preserve himself in flight.

The mule will never go so far as to get hysterical or lose his head. He may be bucking high, wide, and handsome, and making a supreme effort to throw his rider, but in the back of his mind he will always have a caution for his footing and a wary eye for obstacles that he may run into. A horse bucking toward a gully or *barranca,* will be so intent on throwing his rider, he will buck into the gulch and break his neck: the mule never. When in danger, he will forget his fight with his rider and stop meekly only to resume his bucking some other time when his rider least expects it. He may take a trigger-rein and spin, stop or back, putting many horses to shame. But only while he is in the mood, for a reined mule is reined only when he chooses to be so.

He is the only animal the burro will not lead home, for there is kinship, and blood is thicker than water.

Old Men and the Burro . . .

The old men would smile with pleasure and wax eloquent when asked to enumerate the virtues of the faithful and staunch companion of their boyhood. *"Si"* they would say, "the burro is brave. He is the only domestic animal besides the dog who will fight to the bitter end and never quit. The dread puma will run away from a rain of kicks dealt by this sturdy battler."

Sleepy-headed, long-eared, uncouth, mount him and enter a corral with a bull on the prod. You will need no rein to guide him for he can see danger and avoid it without the aid of his rider. The burro may be hooked once but once only. If he lives, no bull will ever strike him again.

Men who have ridden him against fighting bulls say he will dodge and spin away from the thrust of the horns faster than any horse. Burros are put in a herd of horses to protect the foals. The proud stallion may prance about his mares and fight others away in jealousy, but it will be the under-sized alien in the band who will keep his head when the rest are in panic and fight the wolves away from the colts. A burro or mule will kill a wolf.

A horse in close quarters with an enraged bull will often lose his head. The burro will keep his, and coolly avoid the bull's rushes while the horse will become frantic, and as a consequence, lose his life.

All the old-timers agree on the long eared one's bucking ability. They all say that while a burro will throw his rider a shorter distance than a horse he will throw him twice as hard. They all agree that a fall from a burro hurts a good deal more than a fall from a horse. They often used to say that the hardest fall in their lifetime was when they were bucked off a jackass.

Four Old Men . . .

And so we come to the time forty or more years ago when I met the four old men who had come to the end of the trail and waited for death in the Canon de Soledad on the edge of the Mojave desert.

The first old vaqueros I knew sat in the shade of the cottonwoods, rolled brown paper cigarettes, smoked gravely, and told in old voices tales of vaqueros, brave horses, fierce wild cattle, bandits, and buried treasure—Leonardo Ruiz, Dolores Martinez, Jim McFall and Juan Ardia, heroes all to me, a *kid* among old men.

Don Leonardo, the youngest of the four, owned the little ranch along the creek where the old men met to talk. He was a bearded small-featured man with a face like some fine-drawn Spanish vignette, in manner courteous and dignified. His grandfather, a king's soldier, had come by the old Spanish road that crosses the continent from the gulf to the Pacific at San Blas, and up the coast by ship. He had brought a great knowledge of horses and horsemanship, tradition and a love for books. Don Leonardo had inherited much of this, and was the only one of the four who could read and write, a rare accomplishment in those days.

He owned two of the few Spanish horses left alive of the wild desert bands once so numerous; two old zebra-striped mustangs, one a line-backed buckskin, the other a *grullo,* always fat. Small-eared, short-coupled, they stood in a corner of the corral and dozed on three legs. Now and then one would waken, prick one ear forward, then lay it back and doze again.

My First Lesson . . .

One day he found me rubbing the old buckskin with an oiled rag to make him shine, and knew I would never be happy again away from the smell of horses.

And from then on he told me many curious things about them. He took great pains to show the proper way to saddle a horse, which is the method used in Castile, he said, "*Se ensilla, como en Castilla*". First the bridle reins over the neck, then the bridle and bit, the bars of the mouthpiece one inch above the canines in a horse, and if a mare one inch and a half from the last tooth. The two fingers of the hand should fit loosely between the horse's jaw and the chin strap.

After saddling, he had me turn the horse around a couple of times. This *paso de la muerte* was because many horses dislike tight cinches, and the grave is ever yawning for the careless horseman. He gave me one of the great pleasures of my life when he helped me mount for my first horseback ride. Admonishing me to sit and ride like a gentleman, for the rules on riding laid down by Don Quixote in his advice to Sancho are still the standard in horsemanship, and the end to strive for by those who speak the tongue of Cervantes, though the old soldier has been dust these many years.

My First Hero Horse . . .

He told me of the horse's dam, the wild desert mare he had selected because of her stamina and great courage. It had been in Antelope Valley where he and his companions had started the band of mustangs toward the trap they had built for them.

The chase was long and the *puestos* (posted relays) turned them back many times before the weaker ones were roped and tied. Only two mares, a bay *cabos negros* (black points), and a buckskin, were left to be run into the corral far south of the Sierra Pelona. When the captured horses were divided among the hunters he had held out for the bay mare as his share for a good horse is to be esteemed more than life.

A bay or chestnut are the best colors in horses though if an iron-gray is good he is unexcelled. However, when one goes to select a horse he will observe that among the *potros* (colts) there will be one that stands apart from the rest, and because of his speed and spirit will be the leader, and when they run he will take the lead. That horse is the one to select regardless of his color. The mare never became tame. After raising her colt she went back to her *querencia* (home range), but always returned to foal her colts on the ranch of her master. The *grullo* and buckskin were her sons. Horses abounded in those days and could be had for the taking.

So I met my first hero horse, and who would not glory in riding the son of such a mother, and every old rider I've known since then has had a story of some horse he has owned or ridden which was his ideal of equine perfection, and with shining eyes he would recount the many qualities that have been magnified through the years.

Don Leonardo's Stories . . .

Don Leonardo never tired of my questions about horses for were we not kindred souls? Whenever he drove to Mojave or Lancaster in the buckboard he always took me with him, and on the long drive over the desert he told me much of the lore he got from his father, for he had no son of his own to whom he might pass it. Much of this I have forgotten, and this is what has remained in the "nooks and crannies of my recollections."

When the Spaniards drove the Moriscos out of Spain they drove out with them many secrets of Arab horsemanship. Great numbers entered Portugal, where the art of *rejonear,* though practiced in Spain and Latin America, has reached the peak of perfection.

This method is to play the bull on horseback as the animal is played with a cape by a man afoot and not to be confused with the picador, that Judas Iscariot who rides blind-folded nags to their death in the arena.

The Portuguese horseman of today is the master of the Spanish school of equitation, and the reined stock-horse of North and South America is only one phase of the combined knowledge of Arab, Berber and Mameluke.

The Spaniards who came to Mexico during the conquest were of Arab blood, for these new Christians were looked upon with lordly contempt by their conquerors and suffered many indignities, as all vanquished people have through the ages, and so they were eager to embark overseas in an effort to rise above scorn, obloquy and disdain in a new land, where for the lack of good sense they merited defeat, and won victories which they owed to their matchless horses.

The future vaquero had to make only a few adaptations in the Moorish saddle and bridle to develop the equipment in use today. The so-called Chileno or Mexican ring-bit is still the rule in Morocco. It explains how it was possible that the old time riders compel a horse to approach a bear despite his great fear of the animal.

Training a Horse . . .

The Arabs esteem their horses beyond measure of price, treat them with tenderness, and cherish them as a rite and a part of their faith, even as a member of their family. They are trained with much patience and care until the time comes to teach them to fear the spur more than anything in this world, animate or inanimate. This schooling is fraught with so much pain and severity that ever after, at the urge of the spur the horse will throw himself against any living creature or even a stone wall, or in whatever way his master guides. The horse was the major factor in the victories of Islam. The vaquero followed this practice in his mustang and it can be easily understood why old *lazadores* could put a horse, against his instinct and near panic, up to and rope the lord of the North American wilds and truss him up like a pig, and why the *golpe de caballo* (blow with a horse) was so potent a weapon and used by Californians to force their mounts against their opponents and throw them to the ground with a blow of the shoulder.

Don Leonardo on Bits . . .

Don Leonardo was too old to ride for pleasure, but once in a while on a fine Sunday morning he would catch one of the horses, most often the buckskin, and after grooming him and shining the silver on the bit and conchos, put the saddle and bridle on, tie a hair rope with an elaborate knot around his neck, coil the reata in exact dimension, hang it on the saddle and look everything over carefully to see if it was in perfect order. But instead of mounting he would take the slack out of the reins and wrap them around the saddle horn and turn the horse in the yard. The old fellow seemed to know what was expected of him, for like an old charger that smells gunpowder the buckskin, feeble, decrepit, but still splendid, would arch his neck, tuck in his chin, and with ears pricked forward, whirl the cricket in the spade bit in a steady whirr with his tongue that brought a smile to the old man's face, a pleased *"Ah musica,"* and turning to me he would start the day's discourse with a firm *"Un caballo,"* as if encouraged in his views by the horse's elegant behavior. On those days he would take great pains to explain bits and the bitting of horses.

He would smooth out a space on the ground and with a sharp stick draw many different types of mouthpieces explaining the purpose and function of each style, shape or form and the differences in one from the other, and for what kind of mouth it was suited, as horses have hard, soft, or very sensitive organs, and each must be fitted to exact requirements. Then, too, horses differ in the set of the head. No two are alike. They are classified into three types, needing either long, short, straight or curved shank bits. He drew out five different styles of *Chileno,* or Mexican ring bits, each one having some quality different from the others: an Andalusian bit with the bars of the mouthpiece sloping from the cheeks to the center where it arched to bridge the tongue lest it pinch that organ and hurt the horse; one called *"mustache* of the moor" with a curved bar and low port, very mild for young horses in *falsarienda* (*hackamore*) or *dos riendas* (*double reins*); *los barriles* (barrel mouthpiece) with oblong barrels of copper around the steel, called *los sabores* (the flavors) to sweeten a horse's mouth.

One he showed me had what seemed to be scrollwork, and when I asked why it was decorated in such a place he explained that each groove and whorl was no decoration but had the effect of inducing the flow of saliva, for the thing to avoid is a parched mouth. Mouthpieces were made with two, or even three crickets, for playing with the cricket becomes a habit with a horse and he takes the bit more readily. Then, too, the movement of the tongue keeps the mouth from becoming dry and the animal suffers less on a hot day. Half breeds with low or high ports, and plain ports with no crickets but with two or three copper disks for horses with tender mouths were used by old-time vaqueros besides the spade bit, commonly thought nowadays to have been the only type used by California riders. "One must adapt the horse to the bit," he would say, "if he is to learn obedience to the reins, for horses are fierce and proud, but when properly handled are submissive and gentle as lambs, and permit themselves to be dominated by man, and stop, turn and go forward as their masters wish."

Horses' Names . . .

He often recalled some horse he had owned in years gone by, by its color. Vaqueros have a thousand names for the color of a horse, each shade or variation having a distinct and definite term.

It is as a herdsman that the old-timers came closest to their kinsman, the Arab, in naming a horse or a person, *El* something or other for some characteristic or defect in his moral or physical makeup. Quite often much originality was shown in this practice.

Nunca, Nunca . . .

There were a number of things he told me *nunca* (never) to do. Never try to teach a colt something new until he has mastered all of whatever he has been taught.

Never lose sight of a horse's ears, because by their movements he indicates his intentions.

Never put your saddle on a horse that bears the brand of the goat. This brand is easily discerned on the knees of horses that have fallen and are marked for life. They are dangerous to ride. Never water a

The Old Strawberry Roan

horse on a journey without the bit in his mouth. Never lose faith in the endurance of your horse as long as he resists the pull on his tail during a journey. Never discard a horse that is *casquemuleno* (mulefooted).

It is not true, he told me, that the longer a horse is ridden in a hackamore and without a bit the better rein he will have. A horse should have a bit put in his mouth, "when he asks for it," as old-timers say, that is, when he give indications he is ready for it, signs which any old reinsman can comprehend. The best way to avoid bitting a horse when he is changing teeth is to leave off breaking him until he is four years old or more. The ruling factors in bitting an animal are time and patience; time for a horse to make a habit of the bit, and patience to wait until he has that habit.

The horse is at his best after some years of training. His best age is between seven and twelve years, and it is usually a man that is over forty years of age that is an accomplished horseman. For as old vaqueros are wont to say, "the devil is such a devil not because he is the devil but because of the years he spent in practice." Good reined horses and good reinsmen are like him in this respect.

A Man And His Horse . . .

Old-time bosses who knew human nature and applied that knowledge to the handling of horses, cattle, and buckeroos, always gave young riders old horses, and old men young horses, for a youth is combatative and apt to spoil a colt while an experienced rider is tolerant of a green animal when it blunders and will take much time and patience in schooling and making it useful. For unlike other races of men, the *paisanos* accept the horse as they would another *Cristiano;* all its moods and quirks of character are studied and respected as a matter of course. They say of a horse, *"El es asi"* (he is so), and outline minutely every little trait in his makeup, for among them the *equus* is the prince of animals and next to man, seeing that without him no man can approach to that estate which indicates the *caballero.*

Don Leonardo had made a religion of horses and followed it more closely than most disciples follow other creeds. By nature too proud and dignified to show his feelings he nevertheless showed emotion when horses were mentioned and waxed eloquent in speaking of them. The men who lead the strenuous life on the range are often more sentimental than those who walk through life in a city.

There remained with him always the memories of the good horses he had owned and ridden. He often said the horse is the most precious and noble gift a man can bestow or receive. There is nothing that breathes more gentle or bold, more staunch of heart, more loyal of soul than a horse that has a lean head and delicate tapered ears, nor are there eyes on earth as beautiful as those of a horse; soft as those of a woman, that caress when they look, or blaze with fire and courage when danger is near.

Don Leonardo has been sleeping for some years now under an oak tree in San Francisquito Canyon overlooking the old road that leads to the desert. On his monument is a picture of him smiling down on the infrequent traveler that passes by.

The Reined Horse . . .

To Mameluke and Janizary, who fought their battles on horseback with a sword as weapon, and developed the reined horse for close fighting, the high curbed bit which is attributed to them was the most important tool in their equipment. It was developed to keep their high-spirited horses under control when engaged in a duel on horseback. Sword combat required a mount which would respond to the slightest signal from his rider when in the heat and excitement of battle.

From these circumstances the reined horse resulted as did the bits with mouthpieces best suited to him, and of which the spade is a variation. The modern stock horse has as its prototype the Moslem war-horse in this respect. Men who used horses in doing work in which danger was involved, when training a horse for their work, would purposely get the animal into a state of excitement and agitation to teach it while in this condition to obey so that when in a perilous situation the horse would always follow the master's directions blindly and implicity. For a horse will have complete confidence in his rider as long as the master does not transmit his fear to the animal over the oldest medium of communications, the reins. A horse must be angry and on his mettle to do his best work.

The Moon . . .

The moon has much to do with success or failure in breaking horses to rein.

Old-timers say that Luis Lopez, the best reinsman that ever rode on the Tejon, never started to bit up a colt until the moon was full.

Horses bitted when it was tender were invariably hard-mouth runaways.

Vaqueros say of a bad mouth that the horse is *desabrido* (without taste), the result of bitting at the wrong time. The bit should be put in a horse's mouth for the first time when the moon is solid.

They say a horse's nature changes three times from the starting ride up until he is a finished horse—first when he wakes up, forgets his rider and begins to notice other things about him; the second when in the *haquima* he "asks for the bit" pulling on the haquima and foaming of the mouth; third and last, when in the bridle he changes into the character he will have for life.

Enter Dolores Martinez . . .

Don Leonardo had owned the horses and cattle he had worked with all his life, and in this respect he differed from Dolores, the oldest of the four men who had been a rider for wages as a professional vaquero for big cattle companies.

Dolores as a baby had survived the trip over the *camino*, had been a little boy during the gold rush period and panned for the yellow metal with the *gambusinos* in the Sierra Nevada. He had become a vaquero in his teens to go on cattle drives into Utah when the Mormons, the greatest colonists in United States history, had stocked their ranges with California cattle and horses. In Utah were the first markets which opened for animals on overstocked ranges when their only value lay in the hide which covered them. This was long before Belisario Estrada, trail-drive boss, had begun to move cattle out of San Luis over the Temblor range into and across the lower San Joaquin, then over the beetling Sierras through Walkers Pass, or over Greenhorn, into the desert to fight blizzards and storms, or dust and heat. For cattle move slowly, and cattle drives run from one season into another through Nevada and into Utah.

Dolores had amassed little of this world's goods, for the vaquero's trade is seldom productive of wealth since the wage is never princely, and the men of that calling are proverbially free and easy with their money. He had reached a destitute old age. Too old to support himself he was attached to a household but little better off than he was himself. The *paisano* will always share his food, such as it is, though the one who gives and the one who receives may both be left hungry. He never fails to practice his belief that a *platito de frijoles* more or less does not make one any richer or any poorer, inasmuch as he who denies food to his fellow men denies it to God.

He could do but little more than sit in the sun, nevertheless he had not altogether outlived his usefulness. He more than earned his keep by entertaining us with tales of his experiences and passing on to us his accumulated knowledge. Never having known of reference books or libraries, he had not lost the gifts nature bestowed on him. His memory, as is often found in men who cannot read or write, was amazing, and matchless story teller that he was, the most commonplace events in his life in the recountal were full of interest.

With a wealth of wisdom and knowledge stored in his head he would keep us enthralled for hours. He spoke of the great ranches to the north, of great droves of cattle, of *rodeos, corridas* and *lazaderas,* and used terms I understood vaguely and that were not quite clear until years later, when I met the vaquero in the flesh. After I was a little older and had left the lonely canyon to seek the storied land of my boyhood and the demigods of the old man's fancy in the valley of San Joaquin.

He would roll cigarettes and fold up both ends instead of licking the edge of the paper to glue it together, rolling a supply in the morning to last him through the day. Each cigarette was rolled with precise motions as if it were a great breach of rule to deviate from strict form.

In listening to his tales I formed a habit of paying heed to all old men when they talked, and gained a passion for searching through musty old Spanish books in the hope of finding a line or a paragraph that would shed some light on the origin of these men, their animals, and their occupations.

Pico and Murrietta . . .

Don Andres had a fine saddle and bridle, the envy of everyone. When someone ordered a duplicate made, the Indian saddlemaker who made it would go to Don Andres who would give him some money to refuse the order. He wasn't moved by the adage that duplication is flattery. Dolores often described the two *alasanes* (sorrels) hitched to the *caretela* (buckboard) Murrietta drove back here from Sonora many years after the head, supposed to be his, pickled in alcohol, had ceased to be of interest to anyone and was relegated to a shelf in a museum in San Francisco. Manuel Garcia (Murrietta's associate) lived out his old age in Sahuarita, ekeing out a miserable existence milking a few dairy cows, unable to use the gold he had hidden for fear of discovery. But he was found at last and left shot in his rags. The treasure he had lived in misery to protect, still lies buried in Tubutama where Ballesteras, who still lives there, met Murrietta and Garcia with the gold they buried. Near there the warriors of Geronimo had wept over the dead body of Maria Quiros, the girl with the long braids, who cooked for fugitive Apaches escaping from American and Mexican cavalry.

Don Leonardo supplied Dolores with food and lodging, but I soon learned that a present of tobacco from me was sure to bring forth a tale of adventure. During the gold rush days he had carved gold pans out of wood to sell to the miners. He had been a *gambusino* as well as a vaquero. He was called *"el vateaero"* by the other old people.

Utah Cattle Drive . . .

In one of his *cuentos* (tales) he told me of a cattle drive he had made into Utah and since then I have found some other old men in Kern county who made drives into the same country.

Southern California was the big cattle country of the state when Dolores was a young man. He had ridden down to Los Angeles and while there some Mormons boughc a herd of cattle to drive to Utah.

"Dolores was hired as a vaquero with one other *paisano,* the rest of the crew being *Marineros* (sailors) as Americans were often called because they came by ship or jumped ship here on the west coast. The drive was a means of their getting back to the east.

He took six horses of his own, all good bridle horses, though the Mormons had bought a herd of young horses for remounts. The old man told me that the cattle fattened as they progressed but his horses thinned down because, as he asserted, if you wash a horses' back regularly he will get poor and if you don't he will get a sore back.

The Chino . . .

He had been admiring a dark chestnut *chino* (curly haired) mustang all along. Up near what is now Bishop they rested the cattle for some days and built a corral. On the morning the journey was resumed he caught the chino; saddled him in the corral and mounted. He didn't remember how long the horse bucked because after a while he lost consciousness. When he came to one of the sailors was pouring water on him and told him he had found him with his body hanging on the horse's side and both spurs hooked in the cinch.

He knew he wasn't strong enough to ride that horse so he hit on a way to subdue him. Dolores took one of the bars of lead carried for molding bullets, cut out a slug to fit each ear; took two cords, tied two half-hitches around each ear with each cord; then passed them through the hackamore and tied the reins to them.

With this contrivance he kept the horse from bogging his head and in time the animal became gentle enough to ride without them.

I was a small kid then and I've asked a lot of old timers since then if they ever tried that trick and none of them ever had. I asked Adolfo if he ever had and he said not, but informed me the *viejos* (old ones) used many *manas* (stratagems) and were wise when it came to handling horses.

On With the Drive . . .

Well, to get on with our original story; when they finally arrived in Utah he traded his own horses for the *chino* and some boot and with no other companion rode him back to California and settled here in the valley.

Pedro Yorba, a native of Orange county, a member of the Yorba family there, also told me he had been one of the crew of a drive to Utah when he was fifteen years old. When the cattle were delivered he was paid off in gold and rode back to this part of the country with his saddle bag crammed with money he spent in San Luis Obispo.

Don Pedro was a famous roper and his special skill was roping hogs. Years back he trained a grey horse so well he was offered $500 for him, an enormous price for a horse in those days. He wouldn't sell but later gave the horse away. I saw the horse at Granite Station along about 1928. Don Pedro died in 1937 or thereabouts.

Perfecto Cuen, brother of Rafael Cuen, Miller wagon boss, also told me he made drives into eastern Nevada and found the Indians there with documents given by the early Spaniards.

These migrations of cattle account for the slick Visalia saddle-trees and rawhide riatas one finds in use on Nevada and Oregon ranches while other outfits around them will be using different types of equipment.

These drives must antedate the migrations over the Chisholm trail by some years and it must have been in the fifties when the ranges in

Nevada, Oregon and Utah were stocked with the California cattle and mustangs.

Spanish Passports . . .

One of Dolores' favorite tales was of the time when on his way back to California from Utah, he came to an Indian encampment in eastern Nevada where he stopped for the night. The Indians were friendly to men who spoke Spanish and never offered violence or molested the cattle or horses the vaqueros drove on across those states. The Indians still had in their possession documents given them by the first Spaniards to traverse that country and showed them with pride to the vaqueros. They considered themselves allies or subjects of Spain. Dolores said he couldn't read, but understood enough to know that they bore seals and rubricas (curliques) to a signature. One can imagine a scene on the desert along some river, two groups of men staring at a paper both held in reverence, the Indians showing the writing which assured safety to these strangers, and the vaqueros unable to read the words which protected them—both parties paying tribute to a nation long since powerless to enforce its decrees in this new world. It was an experience to hold in their memories and tell to their children.

The old chief of the tribe had questioned him and found Dolores had a knowledge of farming. After he gave a detailed account of sowing, tending and harvesting crops, the old man asked him to stay and take a wife from the tribe and teach his people to farm. This request put Dolores between the devil and the deep sea. He feared to deny the request but he had a bag of gold to spend in Los Angeles. The chief pointed out a girl that he could take for a wife, and later when he was assigned to a tepee, he would not enter it, afraid the girl might follow, so he sat up all night wrapped in his poncho. The next morning they allowed him to leave without interference.

As he rode the women stood in groups about the girl that the chief had offered him, whispering and giggling with many a look in his direction. Though embarrassed he was much relieved, for he knew that had they not wished him to go he never would have escaped, fine horseman though he was. Dolores was only 15 years old when he rode away from the camp with a red face but in the light of experience in later years, he grew to look on his conduct as very manly and grown up during this crisis.

Tailing . . .

The old man would explain every detail in the process of doing each task requiring skill and adoitness which arose in the thousand emergencies in the every-day work of the vaquero.

When branding, cattle should be moving from the roper's right to his left so they will fall on their off side when roped, leaving the near side uppermost for the iron. The trick of throwing a steer by a pull of its tail was discovered by a herdsman of fighting bulls in Spain. One day when in pursuit of a runaway, he overtook the animal on horseback and tried to turn it, but the bull, bred for the arena, was not to be turned by a horse like a common beef. To ride a horse in front of such a bull is to risk having the horse gored. A mother of that game breed that refuses to charge a horse against the pica is unfit to mother fighting bulls and is condemned to the slaughter pens.

Keeping his horse behind the bull for safety's sake, the Spaniard grasped the bull's tail and pulled to change his course. Catching the animal off balance, he threw it, and discovered a new method of handling cattle. *Colear* (to tail), is to turn a running beast head over heels, by riding a horse after the animal, and when the rider is even with the tail, to grasp it, and by pulling throw the animal. This is a very difficult trick which requires a horse with a disposition for the game, one that will close with the steer when necessary, not break or crowd, and obey the reins when the vaquero is ready for the jerk.

Many horses are tried for this work, but few qualify, and those are exceptions, never the rule. A man depends entirely on the horse after reaching the beast. He must lean over, grasp the tail, slide the hand down to its end, throw the leg over it, and holding it between the body of the horse and the leg, then pull when the rear quarters are off the ground. It can be executed and the steer thrown if there is perfect co-ordination between horse and rider, and if the turning is perfect. "But be sure," the old man warned, "the steer isn't carrying burrs or spines in its tail to cut the hand of the *Coleador*."

Tapaderas . . .

Tapaderas, besides protecting the foot from the cold, wet cactus spines and brush, are a shield from many a hurt when one rides in close quarters with cattle in a corral. Bulls and steers will kick at one's foot, when one rides too near. A blow on a member, numb with cold, can be agonizing.

Reatas . . .

When the old-timer needed a reata he first selected a steer of the proper age and condition, and when the moon was in the right sign for starting a rope, slaughtered it and removed the hide with great care. The strength and pliability of the reata depended as much on the animal's condition and on the phase of the moon, as in the braiding which was done in winter when the strands *se prestaban* (were workable), since they were damp and could be pulled tighter when braiding. The strongest ropes were made from hides taken from steers that had died from hunger. These reatas turned black when braided and in use and needed constant tallowing. There is little fat on a beast that starves to death in winter. The best hides came from cattle bred on the coast, there being much less sun and more shade than in desert country, where the hot dry climate burned out the hide.

Cattle were seldom hurt when roped and thrown by an expert *lazador* who knew how to *choriar* (play) the reata, as an expert angler plays the reel and line when a game fish is hooked. The vaquero judged the jerks by letting the turns round the horn slip or hold so the beast would fall softly, as tumblers in a circus roll and fall without getting hurt. Like all artistic souls, the old-timer took great pride in his work, and could lay a full grown bull on the ground as gently as a mother would lay her babe in a cradle. By taking one turn to stop it he could release the turn when the animal was thrown, giving it a chance to break its fall.

The man who hurt an animal when roping it was considered very clumsy and not classed as a roper. For the reata is an intricate tool that is handled best when there is co-ordination of eye and arm. Old timers say, "The hondo commands the loop, the loop the coils, and the throw is decided by the way one whirls and casts."

Dolores told us that in his youth vaqueros received ten cents a head for each calf they caught, branded and marked instead of regular wages. They turned in the tips of the calves' ears when collecting their pay.

Under this arrangement a man became adept at handling the long reata or went hungry.

The Ox . . .

Their skill in roping and tying wild cattle in brush and mountain would have availed the vaqueros nothing in getting them to the home

Perez Threw an Over Handed Loop

corral, had they not had that faithful servant of man from time immemorial, the ox, to aid them. The man who trained and managed the *cabestros* was one of the most important members of the ranch crew in the old days. His title was *caporal de cabestros.* He taught each ox to know its name, to know and keep its position in the band when traveling, to follow its leader, the *capitan,* as the lead steer was called, and when reaching a destination to stand with the rest of the band until some wild steer was roped and tied. Then, when its name was called, it was taught to approach and kneel to be yoked to the hog-tied *bronco* he was to lead home.

Training oxen as leaders is an old skilled art in the peninsula, a profession as old as bull fighting in Spain. It is with *cabestros* (leaders) that fierce bulls are handled and moved. To rope and throw a bull intended for the arena will ruin the animal as a fighter. Each *cabestro* had a hole bored through one horn. On some ranches each vaquero carried a drill of fine steel, the handle of which was the hollow tip of a bull's horn. This handle was filled with tallow to grease the drill when boring a hole in the horn of a wild steer when it was caught. Two steel pins went with each *cabestro,* one pin went into the hole in the ox's horn, the other into that of the wild steer, then they were yoked up and were turned loose, the leader invariably bringing the wild one home to the corral, though it sometimes took several days for the pair to arrive.

Many old vaqueros believe the ox to be a reasoning animal. A *cabestro* will lead any animal whatever the breed, out of the brush, and take care of himself, while the burro will fail when lashed to a wild mule. Not being afraid of it he will not pull it home for someone to relieve him of it. The ox never leaves the road or trail when turned loose with a wild one. The instant the captive heads for the brush, the *cabestro* will stop, and only when it starts in the right direction will the *cabestro* go on again.

Lead oxen had a hole bored in either the right or left horn, never in both. Sometimes after years of wear the hold wore through and another was bored, always in the same horn. An ox will work only on the side he is trained on. They are either near or off side. Whenever available, one was *mancornado* (lashed) to each side of a bull and that gentleman was marched home in record time.

A bull is the most troublesome to lead because he will gore the ox if not lashed horn to horn. The *mancuernas* (thongs) were tied to the ox's horns then around the base of the bull's, and down to a forefoot which prevented him from hooking his *cabestro.*

Cattle were still plentiful when the great fire of 1888 swept through the mountains. Great numbers of the wild-eyed *cimarones* escaped the flames by moving down upon the plains and canyons, where riders seeing them were surprised at the vast amount of cattle those mountains could hold.

Each vaquero was given a number of *ligas* (thongs) a *brasada y media* (roughly nine feet) in length. A *brasada* from *brazos* (arms) is a measure, the length from left to right hand when the arms are spread horizontally, about the height of the person measured. These *ligas* were used to tie the wild cattle when they were caught and left to wait for the oxen to lead them home. The record on the Tejon for roping and tying wild cattle in the mountains was eight head roped and hogtied in one day.

Jim McFall and Tiburcio Vasquez . . .

I have often wondered what storm of life cast the player of bagpipes, Angus McFall, into the red hills of Sonora, and whether he had ever matched his bagpipes with the Yaqui war drums by piping a charge into battle to the accompanying throb of tom-toms. Perhaps he had been shipwrecked on the west coast of Mexico, that graveyard of ships. He found a mate in a daughter of the country, and with her joined in the rush for gold across the desert to meet his death in the Colorado in flood stage. He left a 15-year-old widow with a babe in arms to cross the terrible sand dunes and begin life in a strange land without a protector.

Jim McFall, the babe, when twelve years old ran away from home after his stepfather whipped him and joined a crew of men engaged in driving a band of horses into a middle-western state, probably mounts for Civil War cavalry. He was a tacitum, close-mouthed, old man when I knew him. A ruddy complexion and light brown hair gave him the appearance of a gringo until he swore in Spanish. His mother had been one of the blonde *sonorenas* quite common around Hermosillo and Altar. I always had a great fear of him, for he seemed to be always in a bad humor.

One day carrying the Sunday paper, I stopped at the shack he had built on the forty acres that was his homestead, and sat down to share my treasure with him. I opened the magazine section to show him something about Vasquez, and read to him an account of the bandit's capture. The story made him angry and he said it was not true. In a few terse sentences, he told that though the bandit's captors claimed to have planned his capture and found his hideout after much effort, Vasquez was betrayed into their hands, sold for silver, by a female Judas, consort of Greek George, the camel driver.

Vasquez, who through a strange twist of fate was to know the bitter irony of receiving bouquets from women while lying in jail waiting to be hanged through the treachery of one of the sex, who as old people say *Cambio oro por plata* (exchanged gold for silver). But it brought her no joy, for as the old man said, "Did she not die shortly after screaming through the night of her last agony?" in fear or remorse for breaking faith.

Vasquez, like Murrietta, had the sympathy of the *paisano*, for men from far down in Sonora north to where the last vestiges of Spain end, who knew many things regarding the bandit lived through a lifetime and died without speaking of them.

All the senoras in California have fed Tiburcio Vasquez, every old timer I've ever talked with claims that his mother, grandmother or some female relative cooked a meal for the bandit who must have been always hungry or have made a career of being the man who came to dinner.

McFall alone of all the old men, who claimed to have known Vasquez, knew the woman's name who betrayed him. It was Teresa.

The bandit met her the night she wore hoop skirts at a dance. The sheriff and his deputies, hunting the fugitive, surrounded the dance hall barring all windows and doors. When the sheriff walked in to look over the dancers, Vasquez crawled under Teresa's skirts, which hid him effectively. The sheriff, baffled, left empty handed. She was *muy lunatica* (under the influence of the moon). Having been born under an unfavorable phase, she had spells of madness, old *paisanos* say. It was a malady quite common in the old days. She had a *lunar* (birthmark) under her jaw to warn men of her disposition, but Vasquez feared no man, woman, superstition or devil.

There was a man who owned an old ranch some ten miles south of Bakersfield. *Paisanos* say he was so rich that when approaching some ranch house and the dogs ran out barking at him he would throw a 20 dollar gold piece at them and say, "Here, that your master may buy you meat, so that you need not eat the guest God sends him."

It was on this ranch that Vasquez started his career of banditry, though many people looked upon him as an avenger of the wrongs to the *paisano*. Thus in little bits, up until Jim McFall died in 1916, I got a few facts and tales. The one of the locket and the 50 dollar slug was the most interesting, since I know the woman who has the trinket in her possession today. When Jim was a young man, gold, jewels, money, might be laid anywhere and be safe from thieves, for there were none. Only a *ratero* (petty thief) would stoop to such a theft. But cattle and horses were legitimate prey. To purloin cows or horses was not only considered equitable, but an honorable profession, provided of course the owner was not around. Many times one has heard some saddle-warped old-timer say in giving counsel to his children, "I have always been a *trabajador* (worker), never a *bandido*. I have stolen cattle and horses yes, but that is not dishonor, for everyone did it in the old days."

Jim had made a foray into southern California, and had left hotly pursued by the vaqueros of the man he was trying to relieve of some horses. He at last arrived at the San Gabriel river and stood there trying to decide which death was preferable, hanging or drowning. It was one of those rare years, when the rains are plentiful in California, and the river was flooded.

At this moment a man on a grey horse rode up. He was of a sardonic cast of a countenance, enhanced by a long drooping mustache, sideburns, a hooked nose with nervous flaring nostrils. His eyes continually roved, as if his life depended on his being able to discern everything about him. He looked across and saying, "*Vamos*" rode into the surging water. Jim, deciding death by drowning preferable, plunged after him. They made it across, but lost their horses. One went down in the river, the other ran off when reaching the opposite bank, scared by the men hanging onto his tail.

After some walking they came to a *ranchito* where a grey horse was feeding in a little pasture. The grey had a rope dragging. Catching him, they rode double to a house in Cahuenga Pass where a woman, Teresa, lived. Here Jim found his companion to be Tiburcio Vasquez, and learned that the bandit had been in such a hurry, not because he was pursued like Jim, but because the woman had sent him a message saying she was lonely. It was Vasquez and not the man who was pursued and in danger of hanging that had been the one to lead the way into the flood.

The next day, with a good horse between his knees, the Sierra

Madre behind him, and the refuge in the Tejon's La Trampa in sight, Jim opened the locket and found a fifty dollar slug inside, a gift from the man, who swam torrents to reach a woman.

Jim was a boy then, and all through his life held in hero worship this man, and in his old age was to say a word or two in his defense, little dreaming his hearer would grasp them and remember when the time came years later to set them down in writing.

Tiburcio Vasquez, the bandit, who, like Murrietta has become a legend and the subject of numerous tales, was hanged for murder, though he swore up until he took his last breath, he never killed a man in his life.

This claim of innocence, on the part of the bandit, was owned to be true by all the old timers I knew who had been contemporary with him. One old man, a follower of his, asserted repeatedly that Vasquez was guiltless on that particular charge. That this *viejo* had been a follower of his and knew the case was the positive belief of the *paisanos* old enough to have known the general course of his life.

The old timers say Vasquez always rode a gray horse with a black saddle and a number of them told me they kept one of his grays in a corral for the bandido to change to when he came by escaping from some sheriff or posse. They say his grays were placed in ranches over most of California and he could be sure of a fresh horse at all times.

Vasquez attended dances in Kern county and was like all bandits *muy enamorado,* a great lover. The *tulares* served as a hiding place for him on more than one occasion. So say the old timers.

Kern county, in early days, was plagued with horse thieves and among the tales told of the wily desperadoes is the following *cuento de Vasquez* which I've heard from a number of old timers. It illustrates the legendary cunning of the bandit.

A Bandit's Ruse . . .

The bandit and his crew were in camp in the hills on the edge of Carriso plains and were planning a horse stealing expedition when a disagreement between the men and their leader ended in the leader with one retainer, the only one remaining faithful to him, leaving camp in a huff. One of the mutineers assumed the leadership of the band and proceeded to put their plan into execution.

Their strategy was to steal a few head of horses from each ranch to avoid arousing the suspicions of the ranchers who would not notice the absence of their stock. The bandits had built an enclosure up a secluded canyon and as they brought in the animals a few at a time they were held in this corral. Among the stolen horses were two old, gentle bell mares.

Vasquez, in the meantime hovered near the scene of operations watching their moves from a distance without their knowing it. When all the horses in the surrounding country that could be stolen safely had been brought in and the mutineers were about to move out with the stock to find a market for the animals, Vasquez and his follower rode down to the corral after dark.

The bandits, except for one man on guard, who sat by the fire in camp, had all rolled in their blankets and were asleep. Vasquez and his henchman caught the two bell mares and took the bells off. The leader drove the horses out and his aide tied a rope across one corner of the corral; then hung the bells to this rope and all through the night at intervals gave the rope a shake to simulate the movements of the mares.

The bandit at the fire hearing the occasional sound of the bells thought the horses safe and never moved from the fire. The bell ringer left just before daylight to overtake his chief.

The mutineers on arising in the morning went to the corral to discover the trick played on them by their former leader who, with six hours' start was miles away and had left them to walk many weary steps before finding more horses to ride.

The old timer who told me this story said the bandits were afoot four days before they got mounts on which to leave the country.

Successful Banditry . . .

Out on the plain below San Emidio there stood a lone pepper tree close to the road from Maricopa to Bakersfield. Ramon Dominquez, (Frank's father) with his wife and children, were living near this spot when Vasquez and his band, driving a herd of three hundred mules they had stolen on the coast, stopped and asked Ramon's wife to cook some food for them. After she had fed them they rode on with the stock which they drove into Lower California where they sold them. This was one of the few successful exploits of the hard riding bandit.

Vaqueros de Nombre . . .

Frank Dominquez was born in *Campo del Medio,* now Castro Lane, in 1871 and all the old men who knew him when he was a vaquero say that he was one of the best.

Many of the old timers have passed on over the great divide during the 10 or more years Mr. Dominquez has lingered on a bed of pain paralyzed, unable to move. But the old vaquero survives, a relic of another day, his eyes still bright, his memory clear.

The years he spent breaking horses on Santa Rosa have taken their toll. The jolts and breaks he got in his youth have told on his old body. He is pleased when someone drops in to hear his stories of *vaqueros de nota* (cowboys of note), bandits, and *Hechiseros* (magicians) who rode here in early days.

Manuel Verdugo . . .

He tells of Manuel Verdugo, the *ginete,* a man of note, who when a horse bucked with him would take a short hold on the left rein close to the hacamore, throw his weight on the left stirrup, lift his right leg high, and when the horse's feet were off the ground, Verdugo would give a hard pull on the rein. The horse would fall, the man's boot heel would act as a break when it struck the ground and he would land sitting on the horse with his legs in the clear. Only a very active man and a finished balance rider could ever hope to do this stunt successfully and without being hurt. A rider who clinched with his spurs to ride a bucking horse could never attain such skill.

The Sorrell . . .

Dominquez tells of another *vaquero de nombre,* Juan Perez. For some years a sorrel horse had been running in the mountains along Poso Creek. The horse had escaped from his owner soon after he was broken and had eluded all the riders who had tried to catch him. Dominquez, Bautiste Albritre and Perez were one day riding along the creek when they saw this horse coming down a steep ridge on his way to water. Perez directed the other two to circle back, get behind the outlaw and rush down on him so he would stay on the trail he was following. Perez tied up his horse, took his reata, and went up the trail to a rock which afforded a little space in which to hide.

He knelt crouching behind the rock and made a loop and set it carefully spread out on the ground holding it with four or five coils in his right hand. He waited hoping the horse would be so intent on the riders behind him he would fail to notice the roper along his path.

The two riders got above the horse and rode down on him at a dead run, swinging their ropes. As the animal passed, Perez threw overhanded and caught him around the neck. When the rope tightened the man was jerked forward, but he kept his legs braced rigidly. The horse dragged him sliding until the soles of his boots were worn through. The horse was beginning to choke down when Albitre arrived to put another loop around his neck and hold him with the rope around the horn.

A Vasquez Story . . .

When Dominquez was a small boy his father told him this story: He visited the jail in Los Angeles and saw Vasquez sitting with his shame, thinking long black thoughts that drifted back over many varied scenes and incidents in his life as a fugitive. His memory went back to the time when he was pursued and he hid his panting horse and entered the dance hall, wondering what whim of the devil had caused him to do so.

It is fated that one man is born to riches, the other to poverty and he to be a herder of cattle. Women are loathe to smile on the poor vaquero, for life with him at best is fraught with labor, hardship, and squalor. He had to steal the good gray horses and silvered black saddles he had gloried in riding. For what else is there to enjoy in this life other than good horses? Of course, there was the woman, but she had made the world drop out from under him when she betrayed him to the law in the shack hidden in the Cahuenga hills. There was no hope now; it had died the day she had hidden his gun to render him helpless and impotent. There remained only black despair, for it makes all the difference in life or even death if hope is left or left out.

For of what value to a doomed man are love notes and flowers from women with the milk white skin that fascinates men when it was a black-haired traitress who had caused his world to drop out from under him?

Dressed for the Dance . . .

Dominquez told me that back in the time when Alex Godey owned the San Emidio there was a settlement about seven or eight miles below the ranch. The people of the village would have a dance and vaqueros came from far and near to attend the fiesta dressed in their best, clean working clothes and a silk scarf or a silver-mounted hat band to set off the ensemble, for one can never divorce the rider who calls the golden state his home from the sheen of silver and the shimmer of silk, no matter how poor he may be. In those days a man wearing a Stetson hat and a good pair of boots was considered well dressed.

Joaquin Murrietta . . .

If the *paisanos* from San Diego to Sonoma, who claimed to have known the California bandit Joaquin Murrietta were telling the truth, the *bandido* must have spent all his time here in racing his mustang with the express purpose of introducing himself into every isolated ranch and pueblo in the state, or else have been a perpetual one-man parade with banners, for so many persons to have known him.

The majority of the people who made this claim were not born until after Murrietta had gone back to Sonora to live to a ripe old age and die in bed.

But nothing in California history has left so many phantoms and legends in our hills and sierra as the *bandido*. The *cuentos de Murrietta* have become the folk lore of the *hijos del pais*.

Lover and Digger

The numerous bandit tales told by the *viejos* (old people) all agree on two points, the robber was *muy enamorado* (a great lover) and a champion pick and shovel man. How he and his confederates found enough nights in the short period they were active here to bury all the gold that is said to be in every locality in the state where Spanish speaking people settled is a mystery.

Though the bandits naturally must have used as much secrecy as possible in their nocturnal operations, the *entieros* (buried treasures) have become common knowledge to the *paisanos*.

This inconsistency can be made clear, however, by asking any old vaquero to explain it. Buried gold, the vaquero will say, gives off a heat which breaks into a flame above ground, over the *entiero* and if it is your destiny to be rich and your heart is free of greed and avarice, you will not be afraid to seek the location of the gold when on certain nights—depending on the phase of the moon—you will see a greenish light glowing over the treasure. It is to lead you to wealth. But if your intentions are evil in respect to the future use of the money, you may

Vasquez Hid Under Teresa's Skirts

see the light but it will move away from the *entiero* and lead you into quicksand, over a cliff or into some other dangerous situation.

Noisy Bullion . . .

Buried gold, that is minted money, he will further elucidate, has a means of revealing its proximity by making noises in the night such as groans, the sound of rattling chains or deep rumblings.

Kern county has its share of buried treasure. Old-timers say strange noises are heard at night on most of the old Spanish ranches here. Tecuya is known for sounds of phantom footsteps so regular in their occurrence the people living there moved outside the house in order to sleep. Of late years the footsteps have ceased since a Spaniard herding sheep there found the treasure, or so they say. The deep rumbling on Temblor have been the cause of much searching for treasure in the vicinity of that ranch. Plieto has its peculiar noise, the sounds there are like the rattling of chains.

Then too, there is the spectral rider on the white horse seen on foggy nights on Tejon. This proves, say the *viejos,* the bandits did not overlook Kern county in thier labors, which brings us back to the tale of the bandit, the gold and the lady.

The old-time *malditos* (bad ones) were not always improvident. There was a particular *ladron* (robber), who, having stolen a woman somewhere up the valley had also pirated a large sum of money, so large a sum in fact that it made a load for a pack horse. This man was on his way south with his loot when he entered Alamos creek near Comanche Point.

There the bandit and his lady love were captured by the Indians who held them all night. The next morning they returned the saddle horses. The pair departed, but the natives kept the packhorse and the gold. The Indians at that time, or so the story goes, did not know the use of money, and, thinking the yellow stuff would bring evil upon them, took it up Tunas canyon and poured it into a deep fissure of a cliff.

This *cuento* was one of the buried treasure stories known to the *viejos* on Tejon who are sure the gold has not been found yet because every now and then some old vaquero caught out on the plains after dark would come in and tell of seeing a greenish light glowing off toward El Canon de Las Tunas.

Jose Jesus Lopez . . .

A famous Kern county pioneer was the late J. J. Lopez, superintendent of El Tejon ranch. His forebears superintended the building of San Gabriel and San Fernando missions at the time this country was fighting its war of revolution against the British. His grandparents had been members of the great Anza expedition in to California in 1776. In 1872 he came to the San Joaquin valley and a year later went to work for El Tejon ranch.

Many are the stories about him, although there were *mayordomos* better known among cowboys of that day, none was as much of a study in contrasts as the colorful old-time cattle boss. I remember the day I rode into the Tejon vaquero camp, a kid looking for a job riding.

Nacho, the man in camp, told me to put my horse up and wait for the boss. I had sat down and was fixing something on my saddle when, looking up, I found watching me, an elderly man dressed in a pair of corduroy pants faded white from countless washings, a pair of World War I G.I. canvas leggings, a mohair coat affected by office workers of that period, shoes and an old hat. Surely, I thought, this couldn't be the renowned Don Jose Jesus Lopez. Bosses of his reputation and attainments did not dress like this.

But when he spoke in a chipped nasal Spanish I noted it was correct Castillian and pronounced with the proper accent. His bold, hawk face stamped with the air one finds in men who have many responsibilities, belied the near shabbiness of his clothing. As he looked down at me he gave me the impression of a benevolent bird of prey.

Man of Parts . . .

Well, I got a job and as time passed I found Don Jesus a man of many parts—shrewd, well educated, with a keen sense of humor. He must have found no little enjoyment in outwitting his men because he was always one jump ahead of them. Whenever there was mischief brewing he seemed to sense it as a parent will in a child.

Perhaps it was the involuntary response to courtesy which made every one respect him, for he never forgot to address his men as one Spanish gentleman addresses another. He never raised his voice no matter what the provocation, though when his mustache bristled it was a sign he was angry.

Plenty of Nerve . . .

I was to learn, too, that no one had ever had the temerity to use familiarity with this descendant of the king's soldier. He had plenty of nerve, for there was the horse that bucked him off repeatedly. The sorrel horse, Tomate, was gentle enough under normal circumstances, but whenever Don Jesus, while riding him out on the range would find a *lepe* (orphaned calf) he would be struck with compassion.

"Oh *pobrecito*" (poor little one) he would say, and order one of the men to hand it up and he would then lay it across the saddle in front of him.

After traveling some distance the calf would tire of his uncomfortable position and start struggling, whereupon the horse would buck both calf and rider off. This would happen quite often, for the old man persisted in practicing the virtue of compassion.

Adolfo says it was hard to pull the wool over his eyes. He observed too much at times. One day Adolfo had finished making a riata and had hung it on his saddle. The boss, watching him, remarked that it was a nice rope. He forgot the incident and about two weeks later Don Jesus was standing near him and asked to see the riata which Adolfo handed to him.

"Oh, Adolfo," he said, "you have a singular article here. It stretches itself by hanging on your saddle."

There had been no branding done during the two weeks and the old man knew Adolfo had been roping by moonlight when the bosses were asleep to stretch that riata!

I will always remember Don Jesus with gratitude for the string of horses he gave me were the best ones I ever rode.

Don Jesus once bought a bunch of condemned army horses all sound but hard to handle. When they arrived on the ranch he issued them out to the vaqueros who were young enough to ride tough horses. The vaqueros not wanting to ride them, took leave of absence, each one hoping to stay away long enough for the horses to be distributed before he got back.

But Don Jesus, who could read his *hijitos'* minds since he had raised them, calmly ordered the horses turned out, leaving only a few old ones for the aged men who remained, it being winter and no work other than breaking was going on. Some time later the boys returned, of course broke as usual and glad to work even on hard-mouthed, stampeding army horses. The wise old boss had all the horses waiting for them!

While he worked his vaqueros early and late (the Tejon was famous those days for long hard days and meals that were far apart), no man who ever worked for him will fail to smile when the name of Jose Jesus Lopez is mentioned.

Vaquero Wages . . .

Professional vaqueros rode first for one company, then another. A youngster 12 to 14 years of age, school not being compulsory, received a starting wage of $5 a month. After a year or two it was raised to $12, and it usually took about five years of working experience to earn $25 a month, the standard wage for a rider at that time. Reatas and hair ropes were furnished free to the vaqueros. The ropes were never coiled and strapped to a saddle, but laid ready to hand over the fork loop, ready for instant use. The old-time riders roped so constantly in those days, they developed deep callouses on the edge of their hand from taking the turns around the horn. Twisted hair and rawhide braided equipment was used, chiefly because it could be made by hand and cost the poorly-paid buckaroo only the labor of making it.

Many of the vaqueros on Tejon were indebted to Don Jose Jesus Lopez for their start when mere boys on the road to earning a living. He always called them "my boys" since he had raised them. He was wise to the ways of kids, and his vaqueros would chuckle fondly when telling in later years some story of how the old Don had out-smarted them.

Paul tells me that one day when he was a kid he lost his temper and talked back to the old man, who with his moustache bristling with anger, walked off without a word, leaving Paul surprised and scared at his own temerity. He had time to enlarge in his mind his offense against the old boss when about a week later Don Jesus called him aside and said, "*Hijito* (little son), you lost your temper last week; are you in good humor now?" Paul meekly answered, "Si, Senor Don Jesus." "Bueno," the old man said, "it is time for me to lose mine," and he proceeded to give Paul a severe lecture on the evils of insolence, and the danger of losing teeth when boys are disrespectful to their elders. Paul never talked back to him again.

When there was a mean horse in the corral, which nobody wanted to ride, Mr. Lopez would scheme and scheme on how to foist it upon some rider who did not want it, enjoying the buckaroo's discomfort when told he had to ride the "ornery" creature.

The Canelo . . .

"Virtue is its own reward, but virtue in practice sometimes backfires," Adolfo said.

"For example, there is the case of the *canelo* (strawberry roan). I was a kid then, and loafing around camp one Sunday. Having time on my hands I walked out to the corral to look over the horses that had been penned up for an early start, as the rodeos were to begin Monday morning. There was a very handsome *canelo* in the bunch. Seeing he had a sore back, I caught him and got some tallow and was busy doctoring the horse when Don Jesus, who had been watching me, approached and said, "Ah, Adolfo, doctoring the horse? Very good, I will give you that horse to put in your string."

"Well," Adolfo went on, "I was a little bit surprised. It seemed funny to me such a pretty horse was not in somebody's string already."

That evening when the boys came in I told them Don Jesus had given me the horse. They looked at each other and grinned. One of them said, "He's pretty tough and nobody likes to ride him." "The *caporal* (straw boss)," Don Porfirio said, "Don't worry. When he bucks you off he will not run off with your saddle and tear it up (the dread of most cowboys). When you hit the ground he will come up to you to see if you are hurt and wait for you to get on again. Don Jesus has been waiting for a chance to wish him on somebody for a long time." I then got the whole history of the *canelo* from the boys.

Most ranches have a few spoiled horses that are worthless because of this bucking. They take too much time to handle, and after a man or two is hurt by one of them, they are left unridden and are termed outlaws and not put in anyone's string.

While Frank Martinez was top hand on Tejon there were no outlaws. Shy, quiet, soft-spoken Frank was the last of the old-school vaqueros, men who broke a horse, or broke his or its neck. It was a battle of quirt and spurs, and a man's determination against the horse's strength and courage. He liked to ride spoiled horses, and took pride in making a good horse out of a bad one. The horses he broke were always well-reined and good working horses. I never knew a better vaquero during my stay there, for he rode all the bad ones to the finish, and conquered them all in the end.

Well, to return to the *canelo* Adolfo said, "The *canelo* had been started by a young fellow who had to quit him. He was too much horse for that buckaroo. So Frank, the top hand, had taken him and ridden some of the wickedness out of him, and in the process had made a good vaquero horse, although he never quit bucking altogether.

"The first time he bucked I rode him, but afterwards he bucked me off quite a few times. But true to what Porfirio said, he never ran off with my saddle after bucking me off. He waited till I picked myself up and mounted again. The *canelo* had been broken with blinds, and had become so habituated to them that for as long as I rode him the only way I could mount him without a struggle was to take my handkerchief and stuff it into the cheek of the bridle near his eyes. The creature of habit would then stand quietly for me to mount.

"Though he was a good horse to work on, I never wanted him in my string, nor could I trade him to one of the boys. They knew him too well.

"Finally the fall rodeos started, and another man was hired. Don Porfirio said for each of five of us, to give the new vaquero one horse out of our respective string. This was a chance to get rid of the *canelo,* so I gave him to the new vaquero. Though the *caporal* protested, I insisted he was a good horse, which was true.

"So you see," finished Adolfo, "I had to ride that mean horse a long time, just for letting the boss catch me in an act of pity.

"That horse was a *Barileno* mustang. I never could kill him by hard riding."

The Wild-eyed Steer . . .

We had found the big wild-eyed red steer in the Trampa back in the high, brushy sierra. Its wicked looking horns stood straight up from its head as if designed for the purpose of ripping horses wide open. The brute was at least ten years old.

The vaqueros with hard riding had managed to keep it in the herd until the plains were reached where the task became somewhat easier. As long as the beast stayed in the middle of the herd all was well, but whenever he got out in front or on the edge, he would stampede toward the mountains. If a rider rode in front to turn him he would charge the horse then go on until a little bunch of cattle, cut out of the herd, would be driven toward him to lead him back to the others.

When at last he was in the corral he was parted into the closed pen, where he charged and cleared out two vaqueros when they attempted to drive him up the chute, leaving him in solitary possession.

Don Jesus having watched the whole performance, ordered one of the men to open the gate, and he rode into the pen with the maddened steer. He was riding a dainty little sorrel mare and carred a willow switch. Expecting his horse to be gored and he himself killed, I stood in the next corral and watched the steer charge him. The little mare was beautifully reined and Don Jesus lifted her out of the way, and turning her into the steer she drove her shoulder into its flank crushing the big brute against the bars. The old man then grasped its tail and by keeping the mare against the steer, prevented it from turning and goring. After a few slams against the bars the steer entered the chute to escape this old master horseman who at seventy years of age was still living up to a rule of the west and setting a pattern for all good foremen by doing a very dangerous job himself, never asking his men to do a thing he would not do first.

What has made this experience stay in my memories is that all through the struggle with the steer Don Jesus spoke to the animal in formal style addressing him as *Usted,* and requesting him to cease his rebellious fight, as if he were speaking to some recalcitrant child.

How They Came . . .

Every old vaquero had a tale of how he happened to come to the San Joaquin when the great ranches were forming, and what chain of events or circumstances brought him here. Many came *de huida* (in flight), others to make their homes, and some by mistake. The Tejon's *majordomo* for 30 years landed in the valley through an error in geography on the part of the band of men coming over the Anza trail into California.

Among them was Juan Jose Lopez, a little man who rode a gray horse; also Antonio Leivas, the good *ginete* (rider) who rode a buckskin mule that every little while would stampede off across the desert until stopped by Juan Jose on the gray horse, who would counsel Antonio to hold the brute and not let him tire himself out, "Because any minute the naked ones (Apaches) might swoop down on the party, and the jack would be too tired to escape."

Game Juan Jose, who, when a big man broke and wept from the thirst and hardships, said, "I, who am a dwarf almost, do not cry, but I remember."

They missed the San Fernando Mission where they were to meet its *majordomo* Chico Lopez, and came to San Emidio, where Alex Godey sent them to Jesus Lopez. There they remained as capable and valuable men for the remainder of their lives.

Desert Cattle . . .

Old vaqueros told of cattle on the waterless desert that have learned to lick patiently on the cactus plant at a certain spot until the spines are softened and will not enter the flesh. Then the beast eats the heart, and with its nose pushes over the top to reach the life-giving watery pulp in the center of the plant's trunk. By adapting themselves to arid conditions cattle can go six months without water and thus make use of range that otherwise would be valueless.

The Best Vaqueros . . .

The old men who as boys knew the old-time vaqueros say the old ones were very good. In fact the worst of one generation were far better riders and horsemen than the best of the next. Each generation lost some of the skill up until the vaqueros passed out of the picture.

Of the men who were outstanding and had developed a special skill as reinsmen, bucking horse riders or *lazadores*, the names of Luis Lopez, Esquipula Montana, Juan Leivas, Jose Leivas are still remembered by the old-timers of Bakersfield as the best of their day.

Teofilo Lesalda was another artist with the reata who one day riding in company with a boy, Leonardo Chavarria, the wind in their face, came upon a lion feeding upon the remains of a calf. As he carried the reata always to hand, it was only the work of an instant to whirl and throw, catching the lion around the neck. Then running past a tree, he turned at right angles crushing the lion against the trunk, killing it instantly.

Roping Bears . . .

Four of the best vaqueros had their camp at the Lecheria, and every night they had everything ready for in the morning. When the bawling of frightened cattle awakened them, they would know the great *oso pardo* (grizzly bear) was coming down the *cordones* (ridges) driving the cattle in panic before him. Taking their best horses, with their reatas heavily tallowed for the first thirty feet from the hondo, they rode out and roped the *pardo* and killed it with their knives, a dangerous work.

A bear, catching a horse by the tail, could hold and drag him backward, and the reatas were greased so the bear could not pull it toward him and away from the roper. Tomas Pico, Jose Hinio, Juan Leivas, and Tomas Valenzuela made up the most effective bear-roping team on the Tejon ranch in those days.

Cochito Prieto . . .

The winter had been long, cold and dry.

It was late February before the sun could gather enough strength to warm the earth through the frozen crust that keeps the grass from breaking through. A rain had fallen and at long last a few blades had struggled up into the sunlight, too short yet for the cattle to lick.

Cattle had died by the dozens, but the coyotes, their coats sleek and glossy, had grown fat and lazy. It was the herdsman's anxiety for his charges that had brought Don Jesus Lopez out riding the Tejon Range. With him rode soft-voiced, gray-eyed, ruddy-faced Don Toribio Cordova, the *caporal* who always smiled.

As they rode along counting the piles of bones, once healthy beef, or gazing with horror at the ghosts of cattle that remained to stagger weakly about in a vain search for food. Mr. Lopez remarked on the severity of nature. He was lamenting the loss of so many head, when they rode up on the *cochito prieto* (little black pig) and stopped to watch him.

He was a long snouted, razor-backed fellow, stunted from lack of food—the last survivor of a litter that had perished through the hard season. He grunted feebly and smelled here and there, too weak to root into the hard ground.

Don Jesus' heart went out to the little beast trying so hard to hang on to life. Ah! here, at last, was something that might be saved. He dismounted and caught the pig which was too weak to run. Mr. Lopez was wearing a linen duster of that type worn by motorists of that period. He deposited the porker in one of the pockets and mounted again.

He was riding Tomate, the sorrel horse. Tomate was a frustration to Don Jesus. Though beautifully reined and tough as whang leather, the sorrel would buck and had thrown the old man a number of times. But Don Jesus would never turn him over to a young man who could ride him and "knock his ears down," but doggedly kept him for his own mount and religiously saddled him whenever his

Caballo Pajarero

turn came to ride. Cervantes from that Valhalla where good soldiers go must have often smiled down on Don Jesus.

Well, they rode along until the ungrateful *cochito* finding his quarters cramped began to squeal. Tomate, ever ready to renew his feud with the old man, bucked and threw his rider who landed on the pig and crushed it to death. The old man got up, gazed at the dead animal, mounted again, and rode sadly back to the ranch.

Don Toribio told the men of the incident, but none of them laughed. Perhaps a touch of race down through the centuries had made them kindred souls. They could well understand a tilter at windmills, who, though often quixotic, was never ridiculous. Whatever he did was done *con gracia* (gracefully).

Rattle Snakes . . .

For all the large numbers of rattlesnakes we met in our daily life on the range there was, to my knowledge, no man ever bitten, and proportionally few cattle and horses struck by the reptiles.

If the truth were told, the much feared rattler, the symbol of treachery and dread, was more of a benefactor to man than otherwise. It lived on squirrels and other rodents. These, aside from being destructive to plant life, dug burrows in the ground that were far more dangerous to men and horses than were all the rattlers we met. A running horse that stepped into a squirrel hole was almost sure to break a leg, and if moving fast was sure to throw his rider hard enough to hurt him seriously. In fact, most of the serious injuries to men on ranches were caused by falling horses, and squirrel holes were to be blamed.

Whenever we found a snake on the range we would, of course, kill it by whipping it to death with our ropes, a practice I never liked because the snake could strike the rope and imbed and break off its fangs. This could poison one when the rope slid through the hand when roping.

The phrase "to hunt rattlesnakes" on a horse meant his rider was afraid of him, and hoped the horse would get bit and die so the buckaroo wouldn't have to ride him. There was a big bay horse on San Emidio named Barney. He was pretty tough and one man that rode him had some trouble, so one day he rode him over a snake and Barney was bitten. He survived the poison, though he was short-winded and never had much stamina afterwards.

An English bulldog brought from London by his owner was struck by a snake and the Londoner, a visitor at San Emidio, got a car out of town to take himself and the dog to the veterinary. The Englishman stayed in town until the dog recovered. The animal turned loose on the ranch again went right on killing every snake he met, but became so wary in attacking them he was never bitten again.

Jim Loveall and I were riding up a narrow gully with nigh banks on each side in the hills back of Pleito. Jim was in the lead when a snake on the bank above struck at him, missed and rolled down the bank under my horse. I made frantic efforts to get the animal over it before it could be bitten, which I succeeded in doing.

One time I met with one of those singular occurences which do not happen to a man but once in a lifetime. It was late spring. The Miller and Lux buckaroos had been reduced to seven men. The *rango* (horse wrangler) had gone some weeks and the riders were forced to bring in the horses for the day's work themselves. Each man of the crew had a turn at doing that chore.

One morning we were all asleep at Santa Maria in back of McKittrick in the Temblor Hills. Our bedrolls were spread in a row on the ground. Johnnie Light, whose turn it was to wrangle, brought the horses in and came over to wake us up by whacking each man with his rope. The hard blow would bring us awake instantly.

I awoke when the horses came in and lay there watching as Johnnie hit the first sleeper. He was a red-head whose name I have forgotten. He sat up grumbling. The next man was Ernest Ford. Johnnie raised the rope to strike, then stopped, a look of horror on his face. There was a big rattler asleep on top of Ford's feet! Johnnie started whipping the snake to kill it. I heard the burr of the snake rattling, and with the redhead I ran over to pick up our ropes as there were no sticks, stones or clods to kill it with.

Ernest awoke to find himself surrounded by a combination of dangers—the blows of the hard hemp rope and the rattlesnake. He kicked violently, hoping to knock the reptile off his bed, but instead the snake flew up in the air and landed on his chest still rattling. The buckaroo dove under the blankets and we finished the snake. Ernest along with the snake was well thrashed with three men whipping. When it was removed he got up rubbing himself and grumbling that we needn't have hit so hard. We all laughed.

Nevertheless, the buckaroo when spreading his bedroll would always try to avoid spreading it where there were squirrel holes. Snakes live in these holes and come out at night to hunt, and the outdoor sleeper had a dread of the snake crawling into bed with him. The thought of one rolling over in his sleep and on top of a snake would bring a few sleepless nights when the rodeos started, but after a few days the buckaroo would be so tired he would go to sleep anywhere and never worry about snakes or anything else.

One time we found at Painted Rock pump a Garcia steer walking round with a snake hanging from its nose. The snake's fangs had become imbedded in the flesh like fish hooks and the animal's face had swollen until its eyes were closed.

Horses are most often indifferent to rattlesnakes and will walk over them as they lie in the trail. Dogs that are bitten develop a great hatred for them, and nothing will stop them from attacking and killing every snake they find. They catch them in their teeth and snap them to pieces by a violent shake of the head.

One moonlit night we were moving beef to the shipping pen at Caliente. We were on the plains a short way from the vaquero camp when every little while we would hear a snake rattle. The cattle would spread out to avoid it and we would kill it. Sutah Coway, Willie O'Brien, Dave Alexander and I killed twenty-two of the snakes in the space of about a mile. We finally got tired of getting off our horses to kill them and let them go. There must have been hundreds. We thought they were migrating.

Buckaroos would sometimes take a snake skin and stretch it over the top of the cantle. The skin wore for a surprising length of time. The practice of cutting the rattles off and sticking them in the hatband was frowned upon by the old-timers. They said the dust off the rattle, falling in the eyes, would make one blind.

The person to get the biggest scare from a snake, to my knowledge was an Indian kid who rode a horse that was *muy pajarero* (bad to shy). Half blind or near-sighted horses have this habit in the extreme. This boy was riding his horse at a slapping trot and standing up in his stirrups. They came on the snake at a bend in the road. The snake had coiled at the sound of the horse's hoofs and began rattling when the horse made the turn and came upon it. The horse whirled, throwing the kid on the ground, almost on top of the snake.

He rolled away before it could reach him and got up without being bitten. From that day on he never rode that horse again because he said the horse hated and wanted to kill him. *Paisanos* believe that a horse forms a liking or dislike for a person at the first meeting and will never work for the person he does not like. We told the kid that he fell off because he was riding at a trot, which he should never do. A saddle horse either walks or runs and never should be trotted.

In a Barley Field . . .

One day Cheno (Pomuseno) was riding a spooky colt through a barley field on Tejon. Cleve Helm and a kid called Lompoc were riding behind him. The colt was inclined to buck and Cheno was having a hard time keeping him from bogging his head when the barley beards in the field they were riding through tickled his legs. Cleve and Lompoc were enjoying Cheno's predicament immensely. When reaching the middle of the field the colt began to get used to the beards and quit humping his back. They, wishing to prolong the show, gathered up a number of barley heads, and using them as darts, began throwing them at the colt. Cheno would start to turn to see who was throwing when the colt would jump and he would be compelled to give his attention to the horse. After the animal calmed he would start to turn his head again to see who was behind him, when one or the other would throw another barley head, thus it went across the whole section of barley, the colt would be hit, he would hold him up and quiet him, then try to turn to see who his tormentors were, when another head would hit and the performance would be repeated. At long last the field was crossed without poor Cheno ever knowing who the rascal was, that was trying to get him bucked off of his horse.

The Far Away Look . . .

The older buckaroos were sometimes patronizing. With a superior air they would tell us youngsters what an animal would do under certain circumstances and what we should do when confronted with a problem or emergency.

One night we were sitting around the stove in the bunk house on San Emidio and one old-timer was telling us that a wild steer on the prod when roped would charge every horse around, but never the horse that was holding him. One of the kids asked what could a man do if a steer on the prod were roped and thrown and the horse slackened the rope and let the steer up when the rider had dismounted to tie it. The old man answered that it was easy to grab his tail and hang on.

"He will get tired of trying to hook after awhile, and will stop and begin staring at something in the distance. The steer will get a far-away look in his eye and presently run off."

A few days later we found a buckaroo down on the flats below Santiaga, hanging onto a steer's tail. He had been there for some hours and was panting with exhaustion. The steer must have forgotten to get that "faraway look" in his eyes!

Brujeria . . .

Sometimes when I had given him tobacco and no one else was around, Juan would tell me of mountain warriors whose cradle songs boasted the Yaqui knew how to die; learning when life was beginning to someday meet death against an adobe wall defiantly, unflinchingly, chanting the death song of an unconquerable Indian, *El Yaqui que sabe morir,* and of Perieda, the butcher, more bloodthirsty than a lion or tiger, the slayer of Yaquis. Juan's father who had escaped the "butcher of Sonora," told him many tales when he was a boy on long winter nights in the Tehachapi.

A mixture of Yaqui and Piute, secretive, aloof, unfriendly, he talked to me more than to anyone, for as he said, I was the only one who cared whether he smoked or not. He told me stories, and believed them himself, of men possessed of the devil (*diableros*), of witchcraft (*brujeria*), ghosts (*espantos*) and of Patagon (Bigfoot) who was a diablero, because he could not ride bucking horses. Every fall when the rodeos were over the colts were brought in to be broken through the winter. Patagon would immediately get sick and stay so until the colts were all broken and gentle. Everyone knew it was fear which made him ill.

On the same ranch was Diego, called Diego Vaquero, because of his skill. People said he employed black magic in his work. Patagon gave him two reatas as the price for telling him how to stay on a bucking horse. He was told to take two pieces of wood, carve them into the shape of a man's legs and tie one on each side of the saddle. Then when he rode a colt, to curse the devil while the horse bucked. By following these instructions he would be a great rider and no horse would throw him. But if he were to cross himself, the charm would be useless because from then on he belonged to the devil.

Patagon used to put his saddle on a bench and offer 50 cents to any kid on the ranch who could ride it. All the boys tried, but the saddle bucked them off, one after the other. He would lay it on the ground and it would rise and buck as if on an invisible horse. He would turn himself into a bull and his wife into a heifer and together they would roam with the cattle in the mountains. One day the vaqueros roped him when in the form of a bull. They made a fire to heat the irons, and the *caporal* was cutting the mark in his ear when he spoke and told them he was Patagon. The skeptical vaqueros cut a small notch to see if it would show when he assumed human form. The next morning he sat at breakfast and the vaqueros noted the notch in his ear.

One time he left the home ranch for a job in a strange place. The vaquero boss there gave him a mule. The first morning the mule bucked and he whipped and spurred the animal until it quit and rode out with the men to hunt wild cattle in the brush. He came upon two bulls which charged in. He roped and hogtied them after whipping them soundly, and left them for the oxen to lead home. The next day the foreman and two vaqueros were laid up with bruises; they had been the mule and the two bulls Patagon had handled so roughly.

Juan said that one day he was herding a bunch of goats. These were kept on the ranch because they keep disease away from the horses. As long as goats are about, horses will stay in good health. He was near the Indian village and was watching when an Indian woman came out of one of the adobes and walked to a heap of ashes where she disrobed, lay down and rolled in them. Presently she rōse on all fours, shook herself and trotted off. She had assumed the body of a coyote. She came back later carrying a rabbit in her mouth. Juan swore that her family had rabbit for supper that night.

Another time the people living in that cluster of adobes were sitting in the moonlight when one of the women disappeared, and an owl flew over, landing in a tree. The children attacked it with stones, breaking a wing, and it hobbled away. The next morning the news was about that the woman who had disappeared had a broken arm.

Llano de Beslin . . .

Most vaqueros had heard and often spoke of the Llano de Beslin. In the middle of this plain lay a city covered over by a great salt lake. The people of this city did not perish in the flood, but lived on under the waters which covered them. When one is embarked upon it, a hundred or so meters from the shore, one begins to hear the sound of dancing feet, the music of the song, and the jingle of spurs on stamping feet coming from beneath. There is much of *brujeria* (witchery) there. The lake is thick with salt, but a stream of fresh, sweet water jets like a fountain from its center.

Here on this plain is the cave where those wishing to make a compact with the devil are tested. Through it they must pass between rows of fierce wild beasts without fear or faltering. These are stationed there to try their courage. For if those who follow in the path of righteousness must be brave, those who follow the evil one must be doubly so, for all old timers say, "El Diablo hates the whine of a coward."

Forty Years Ago . . .

Back in the sixties, when the southern San Joaquin and the mountains of Kern county teemed with thousands of wild cattle and horses, the great ranch companies began to replace the rancheros of the old Spanish grants. Sonorenas coming into the valley made up the vaquero crews on the great holdings of Miller and Lux and the Kern County Land Company which is still the greatest cattle outfit in North America.

The vaquero of 40 years ago, when I first knew him, was at his best doing his daily work. The ranchers that employed him credit the old timer with absolute devotion to the man who paid his wages and in the majority of cases this was true, though this child of the range required handling with kid gloves, so to speak, in order to get the best out of him, for his pride was easily hurt and being a natural gentleman himself, he demanded gentlemanly treatment in voice and manner. With all this he was as hardboiled as he was soft-hearted.

Profoundly ignorant of everything outside of his own calling and being a past master of the delicate art of ridicule, he laughed disdainfully at a greenhorn's stupidity with a primitive sense of humor somewhat peculiar to men of the range.

Like all children of nature, the vaquero was sentimental and the guitar and *cuento* played a large part in an otherwise drab life. Perhaps the oldtimer's worst trait was professional jealousy, as vaqueros almost always envied each other's skill.

This skill had been passing but there were still a few men left in the Southern San Joaquin Valley who could throw a rope with such artistry that the loop seemed to be guided by an invisible hand to the horns or hoofs it was destined to encircle. These men could hogtie a wild steer or tail a fighting bull and send him head-over-heels to the ground and into submission and ride any of the thousands of sand-locoed horses then ranging the valley.

It was mastery of such feats that made up the entire course of study in a *paisano's* education and in those days the buckeroo was held in high esteem. Many cattle bosses in charge of thousands of head of cattle could not write their own names.

They jogged along through an era and into oblivion, always astride a gelding, for the vaquero had his pride and never rode a mare. Bosses knew that if they wanted a man to quit all they had to do was to give him a mare to put in his string and the buckeroo would roll up his bed and leave the ranch.

Equipment . . .

Shotgun and woolly chaps were the style of the time and flat-brimmed, stiff sombreros were tied down with a ribbon under the chin. Most of the cow men affected silk handkerchiefs worn around the neck, and as much silver mounting on bits and spurs as they could afford. It was a poor vaquero indeed, and one entirely without pride, who did not have some things made of silver about him or his horse.

Each rider owned a trigger-reigned, cricket whirring horse that could turn on a dime, his most prized possession.

A vest was standard equipment as well as the horse-hair *cabestro,* rawhide riata and fringed buckskin gloves. Those who wore denim overalls cunningly rolled up the bottoms of their levis so the fine cloth underneath could be admired. When forced to walk about the vaquero swaggered a little and with reason, for nowhere in the history of the world has any calling been more glorified, more words written about or had more imitators than that of this common laborer and herder of cattle.

Sheepherders . . .

After some months spent on the range, the cowboy would get a yearning for an ordered life and wish for such things as baths and clean shirts. The younger members of the crew would get *ringy* and would remark that a cowboy was a sheepherder with his brains knocked out, because the sheepherder got more pay, better food and was supplied with a tent, whereas the cowboy, when away from the bunkhouse, was sadly out of luck and slept out in the cold.

A buckeroo was always glad of a chance to eat ham and eggs at a sheep camp. He never found them on the chuckwagon. The herders were even supplied with wine.

Frank Martinez, a very blunt man, once walked into Towne's store in Tehachapi and said, "My stomach asked me for something sour," which meant his stomach had rebelled at the monotonous diet of a cow camp and needed a change.

When in town the buckeroo would stuff himself on pie a la mode. One time six riders from Buttonwillow were in town, and as long as their money lasted they took turns standing treat for the pie and ice cream throughout each day. It was in the fall when most of the buckaroos would quit riding and say it was for good, and would sometimes go so far as to sell their saddles. But when spring came around they invariably would come back to the ranch and start riding again.

One cold December day we stood in the rain huddled around a fire at Pimental ranch on the Carriza Plains. The cook was trying to make the smoldering fire blaze, when some wit in the crew went around and asked each one of us if we were crazy. Every one, even the cook, admitted being mad, for as one said, "We must be. If we weren't, we wouldn't be riding in the rain for $40 a month."

The buckaroo is an old-fashioned sort of guy who is always careful to lift his hat clear of his head when he meets a lady. All women are ladies to him.

The vaqueros, when they sat by the campfire on the range, would brag of their conquests among the fair sex, and when they came to town they would stand on the sidewalks and in conversation ride bucking horses, but the fact of the matter is that the buckaroo in reality was as shy as a daisy, and when a woman did accept him he was filled with a surprised awe and marveled at his great good fortune.

The Dude vs. The Vaquero . . .

Try as he might to resemble the true buckaroo, the ubiquitous dude always fails in the end, and that is because his hands are too white, his shirt of a louder pattern, and his attire altogether more gaudy and elegant and much less practical than that of the real cattleman. Then too, the dude lacks that air of free and easy confidence in himself that is the peculiar mark of the vaquero. The dude dares not assume it.

It is when the vaqueros were gathered around the campfire on the range when the days work was done and the weary riders sat or lay at their ease with the firelight casting fitful shadows over their recum-

Rodeo

bent figures, that one had the opportunity to study their character from outward appearances. One was struck with the resemblance in their physical makeup.

Though their features did not resemble each other there was an unmistakable similarity in the expressions on their faces and bodies in general, no doubt, the mark is made by like occupations and experiences. They were then drawn together by the accidents of life or the instinct of adventure, all more or less strange individuals. The flames lighted up faces with beaked aquiline noses that curved into a hook when they grinned sardonically at some remark made by one of their number. I can see them now; long drooping moustaches on lean bold faces. Wearing Pablano hats with stiff brims and flat crowns, with silk scarfs and vests, garter arm bands, overalls over fine cloth pants and tight boots on feet that seemed to be much smaller than those of men who must walk a lot to earn their living.

The Novice . . .

Sometimes a kid that is clumsily courageous and ignorant of the work will amuse the older men by his intensity and keenness for the vaqueros trade. But he will awe them in the end when they see the passionate energy with which he gives himself to the work. If they know his people, and he is of vaqureo stock, they will say it is "in his blood," and that excuses everything. Kids hear old men talk and make an honest effort to maintain the spirit and traditions of the California style of riding, roping and working cattle.

Barbed Wire . . .

There is nothing in the west that is looked upon with more hatred and opprobrium than that product of the devil, the treacherous barbed wire. No one has been cursed as often nor with more true feeling than the man who invented it. Barbed wire, the vaqueros' worst enemy, has ruined more horses than all the fires, floods, drouths, bad handling and acts of God combined. It has caused more disappointment and accidents than any other plague visited upon the inhabitant of the range country. It seems that only the finest, most promising colts are crippled and slashed to ribbons in its barbs. Many bleed to death long before they are found and saved for a worthless existence.

With barbed wire came civilization and its host of farmers, traders, married men with large families, which in time covered the open range land. They plowed the first furrow and no doubt, called it progress to break the sod and ruin the world's finest natural pasture and make an end to the breeding of beef cattle on the open range.

The saga of the vaquero from this point onward has a sad ending. Sadder even than the butchery and extenction of the buffalo and the other myriads of wild life.

The Chuck Wagon . . .

Some time back a friend gave me a print of a group of men riding in a chuck wagon to add to my collection of old pictures. We had been looking through the tattered old album with its old fashioned snapshots. Another was of a girl in a Spanish costume, high comb over bobbed hair, with a yellowed lace mantilla. Still another was the head of an Arabian horse with great mournful eyes and fine small ears, a star in his forehead. She showed me others, pictures of mounted men on shaggy-legged horses wearing silver-mounted bits and flower-stamped saddles with the indispensible rawhide riata.

The picture of the chuck wagon sends one's thoughts drifting back over many varied scenes and experiences of years ago when the old vehicle was still the most important implement in the operation of the vast Buttonwillow Ranch. Though it has long since been discarded and left to dry rot under a gum tree, it is still cherished by its present owner, a son of pioneers who values it for its old associations and as a relic of the day when cattle, vaqueros, and open range were of importance in Kern county.

The men in the picture are dressed in business suits but wear towering old style ten-gallon hats resurrected for this occasion, keep-sakes of happier, able-bodied days when this ride in a parade would not have seemed a whim of decrepitude. Some of them have been broken with poverty, pain, old age and a long course of accidents.

The big, once lusty blond man in the gray suit grips the side of the wagon to steady the chair he sits on. He is afraid. Always in the back of his mind is the memory of the suffering he met the day he saddled the big sorrel horse. He had put his foot in the stirrup and was pulling himself into the saddle when the horse fell over backwards. Cinch binders are tricky brutes. He was taken to the hospital with a broken hip, and when he was put in the cast they told him he would never ride again. He had been luckier, perhaps than Chico Ruiz. Chico had been killed when his horse fell over with him. The old man had been doomed to the rocking chair on the porch until this celebration and had come along to break the monotony of a lame man's existence.

He had once been in charge of this wagon when as straw boss the vaqueros had been under his orders. He remembered the succession of cooks that had come and gone; all bad, some worse than bad if such a condition can be imagined. The food had been vile, *boludo* (lump jaw) beef, beans, rice, macaroni, cooked in black pots over an open fire, and biscuits from a dutch oven. Fried meat, beans, potatoes, greasy food, greasy dishes, flies, muddy water, ashes, and just plain dirt, the smell of cattle over all.

The riders came late in the day half-starved and ate too much. They would go to sleep to awaken sick at their stomachs. At breakfast they would try to wash out the bad taste in their mouths with coffee.

The wagon was pulled by mules and the chief requirements in a cook was that he be able to drive them. His skill in cooking was not even considered. They were terrible, untidy, *belly-robbing,* incompetents. Old Lupe was the cleanest and best of them all. Soup was the dirtiest. Lupe would make what he called stew out of the tenderloins, and steak of the shoulder and chuck. But he often made *pililis* (biscuit dough fried in deep lard like doughnuts) much relished by vaqueros, who being young, had cast-iron stomachs and could digest anything.

The company furnished canned vegetables. The cans were opened together and poured into a pot to make a sort of mulligan. One of the utensils always found on the wagon was a stick six or seven inches long with a rag wound around one end of it and tied with a string. The length of the stick depended on the degree of aversion the cook had for water. It was used for the purpose of washing dishes without danger of getting the hands wet.

The hot cake batter was mixed in a bowl that had never been washed since the first time it was put to use. Each succeeding batch of hot cakes had left a layor of dough to dry and harden and as time passed the space in the bowl became smaller and smaller until it was so small the bowl had to be discarded. The flour had hardened and could not be chipped out without breaking the bowl.

Miller's Hash . . .

We'll make a dash for Miller's hash
On the banks of the San Joaquin
You take the beans, Bill; I'll take the stew
And we'll show these 'Dirty Platers'
What us tule boys can do."

The above remnant is all I remember of the song which had its origin in the San Joaquin Valley in the days of the great ranches and cattle kings. It was sung by the *bindle stiffs* who likewise were born of that era and who half-cynically, half-humorously made the *Dirty Plate Route* known in all the hobo jungles of the United States.

The Dirty Plate Route was the line of ranches and cow camps extending the length of the valley. Henry Miller was afraid of having his numerous haystacks burned by cigarette-smoking sleepers and, as a consequence, gave orders that all men who came to any of his ranches were to fed and lodged whether there was work for them or not.

The result of this order was that Chinese cooks on the various places would often find that at mealtime there would be more guests than workers to be fed. If too many showed up to eat, the flunky would not wash the dishes, but would leave the unwashed plates for the stiffs to eat from, if they ate at all—hence the name Dirty Plate Route.

Whenever an Oriental was faced with the problem of feeding more men than he had prepared food for, he would set all the food out on the table and ring the bell. The dinning room doors were always kept barred until the last notes were rung. The flunky would then throw them open and get out of the way to avoid the rush of the men to the table.

Old-timers who were wise to the game of grab or be left dinnerless, would hurriedly take a seat and fill their plate with what ever was in front of them. One man got the stew, further on another got the beans, and so on. If one was a trifle slow he got little or nothing. A newcomer to the Dirty Plate Route would, after a few lean days, learn to make a dash for Miller's Hash like any old time stiff.

Miller, more than any of the big early-day ranchmen, was responsible for the creation of that unique character which is seldom found nowadays, the California *bindle stiff*. Each stiff owned a *bindle* or *balloon* (bedroll), which they carried on their backs from ranch to ranch. It contained all their worldly possessions. They were the least encumbered of the men who made up the crews on the great ranches of that free-and-easy period.

The buckeroo by contrast had to own his saddle, various bridles, and other tools besides his bedroll, for he never called his blankets and their canvas cover a bindle. He was more heavily burdened than the stiff and couldn't carry them had he wanted to. His boots were not made to walk in.

The stiffs would find work on one ranch or another, as men were continually coming and going even in winter when great numbers of men drifted to California to avoid the rigorous weather in other states.

They could follow the Dirty Plate Route from Bakersfield to near San Francisco, as the ranches or camps were within easy walking distance from one to another the entire length of the great valley.

There were several dining rooms on each headquarters ranch like Buttonwillow, and each man's status was determined by where he took his meals. The office crew ate in what the stiffs enviously called the *blue room*. The foremen and men who had a skilled trade like blacksmith, harness maker, etc., ate in another. The vaqueros luckily had a cook and a camp of their own usually some distance from the men who worked afoot. They stayed pretty much to themselves.

The early day bindle stiff was usually a single man who had wandered here from some eastern state and it was not hard to find men with good educations among them. In those days one often heard Greeley's "Go West, young man," quoted by some cynic sitting on a straw-filled bunk in a room reeking of tobacco smoke and the unwashed bodies of men all blessed or cursed with the itching foot.

Frontier Justice . . .

The old cottonwood tree on Castro Lane is still alive, though some years back a grass fire burned the limb from which hangs this tale. Tomas Castro, with his family and the ox, the only animal he owned when he left Sonora, to start life anew in the San Joaquin, had built a house of tules with a steer hide for a door. It was built in Campo Del Medio (Middle Camp) so named by the men who drove the big herds which rested here on the drive from San Luis Obispo through Temblor over the Sierra into Nevada. It was here that Jim Crocker, driving twenty-five hundred head to a spot east of Tonopah, stopped his herd and gave a little white orphaned calf struggling to keep up with the rest to a little boy. He never dreamed this starving lepe would be the start in life for the barefooted kid.

One day seven men riding gaunt, jaded horses came to the tule house and asked Castro to sell them eggs which he did. The seven rode away into some tules where they had a camp. A short time later a group of well-armed riders with rifles across their saddles rode up and asked if any vaqueros with a band of horses had passed that morning. Castro pointed to the ground saying, "There are tracks to lead you. Follow them."

The men, about fifteen in number, rode away. Some hours later the leader and four others returned and asked Mr. Castro to go with them to where they had left their prisoners and act as a judge, as they were to try and hang them. He protested, but to no avail; the men were armed so he went. The seven who had bought eggs that morning had stolen a band of horses below Los Angeles, at or near the Rancho La Bayona. This posse had been following them closely and had captured them while they were eating in the tules.

The trial was short and swift, such as it was, and immediate preparations were made to hang the culprits. Mr. Castro's two sons Domitildo and Leonides were asked to be witnesses that the thieves got a fair trial!

With the captives was a young boy about seventeen years of age. The rancher was stuck with the lad's appearance and asked him how he happened to be in such company. The boy said he had been walking along the road near Rose Station carrying his *mochila* (pack) on his back when overtaken by the horse-thieves who had offered him a ride. Making a pad of his blanket and thanking his lucky stars he had mounted one of the horses and had come along with the bandits. But now they were to hang him, which would not be so bad if he were an old man, but he was young with much to live for.

Tomas Castro having in mind the age-old sanctuary of the church, took a black blanket and covered the boy with it, claiming immunity for him. But the posse leader contended the boy was guilty, because as he explained if he, Castro, were to rope a steer by the horns and his son should come along and rope it by the hind legs, he would be equally guilty. The boy had ridden a stolen horse so must hang.

Wherewith the captives were hanged to the cottonwood tree, the boy first despite Castro's pleas, for life was cheap in those days to one like the posse leader with murder in his heart.

Don Tomas had rosaries and masses said for the boy, whose name he never knew, for he was a devout man and had sons of his own. Often in later years he would wonder if the boy's mother still waited for him, hanged by men who did not care enough to learn his name.

The stolen horses were sore-footed. Five of the mares could not travel at all, and the owner offered them as a gift to Castro, who refused them in horror. They were left there anyway when the ones able to travel were driven back to their home range. Canducci Castro remembers many things regarding them. In time they had colts which were the start of bands of horses his people were to own. When bred to

Jewett's Hambletonian they produced colts which made their owners famous as breeders of good horses, though mean and hard buckers. One mare which had a spot on her side threw all her colts marked like her. They were such good buckers that one threw Esquepula Montana, who in the opinion of all the old-timers, was the best rider that ever lived. This horse reached the rare age of thirty years. Every July the animal would leave the ranch on Greenhorn and go back to the home place in the valley. Another descendant of the stolen mustangs would return from the mountains in the fall.

Once he was loaned to a man to ride to the mines in Nevada. This man, Refugio, left him in Bishop in a pasture. The horse got out and came back home to Castro Lane. Several people who knew the horse, recognized him on the road miles away from home. One saw him going through Weldon. He went to the ranch on Breckenridge, and finding no one there, went on to the valley, where he was found at the gate one morning waiting for someone to open it.

This horse was fast. Harrison Jones raced a bay against him, and with Vicente Contreras as jockey. The sorrel won.

Cattle Theives . . .

Cattle stealing was encourage by some big cattle owners in those days because they could buy the stolen cattle cheaply.

Joe Salines was foreman for Alex Godey when Montana was jailed for stealing; the big cattlemen he had served got him out of jail and gave him a fast horse. He went back to Sonora where he met death at the hands of a widow whose husband he had shot.

There were many *ladrones* and *coyotes* (robbers and thieves), Castro knew by name, who stole here and drove to San Luis. or stole there and drove here. He remembered their names, but they are all dead now and it is not good to stir up those who are at rest.

Vaqueros would come in and tie their horses to hitching racks and go into a saloon. The *coyotes* (thieves) would ride the horses to Sumner (East Bakersfield) and there pawn them for $10 or $15 at a livery stable, steal another and ride it back to Bakersfield and pawn it. Then there was the big cattleman who would buy a small band in San Diego and drive it to San Francisco. When he got there the herd would have grown to hundreds.

Jim Waldon tells how the ranchers in the valley solved the problem of cattle thieves. Fuentes, a man who had won many races and was a very good rider was hired. Then each rancher contributed his best horse to a pool of horses which were left, one on each ranch between San Luis and Bakersfield. Fuentes was instructed to go to San Luis and stay there until some rancher came in to get help when his horses were stolen. Fuentes kept his horse saddled day and night, and when he got the first news of the thieves, he would head for Bakersfield. With the relays along the route, he would make it into Bakersfield two days before the horse thieves who were met by a posse and hanged. This stopped the large scale thieving in Kern county.

Jose Salinas was the genius who took the Miller and Lux brand and made it into a windowpane, a piece of artistry for which he went to jail, but he was helped out and escaped justice.

Miller's Driver . . .

A. T. Lopez, an old Miller and Lux buckeroo, drove the team for Henry Miller, and like every other man who has known and worked for the cattle king, has an anecdote to add to the numerous legends of Miller that shed light on the old man's singular character. Miller traveled in a buggy or buckboard and covered all the millions of acres under his control, trying to see the condition of every cow, horse, sheep, pig or chicken on his enormous domain. When arriving at a ranch house he would ask the superintendent or foreman in which sections or pastures the stock could be found, and after inspecting them would drive on to the next ranch in his endless rounds of the greatest cattle outfit in North America.

He asked his driver never to open conversation with him, for if he did so he might break some deep train of important thought. Once when Lopez was driving him along a road bordered by a hedge they came to a closed gate. A man trimming the hedge was standing near and the driver called to him to open it. The man turned and said, "Go to the devil. I work for Miller and Lux," and went on with his work. The cattle king opened the gate while Lopez drove through, and that evening Miller asked the foreman the name of the man trimming the hedge and ordered his wages raised ten dollars a month for doing such a good job of work for Miller and Lux.

There are many arguments among old-timers who rode the San Joaquin ranges at to whether Tiburcio Vasquez ever robbed Miller. Most of them contend it was some other bandit and not Vasquez, and that he once made plans to do so but Miller outwitted him. Some claim that Vasquez helped Miller to acquire cattle. Many times they say when driving stolen cattle along near the cattle king's domain, when closely pursued by the sheriff, Vasquez would drive the stock onto Miller's land and make his escape. The cattle in time would increase, and since the brands had been altered or blotted, the cattle eventually came into the possession of Miller through sheriff's sale or otherwise.

The last time Lopez saw Miller was on his last visit to Bloomfield, and he rode in an automobile. There were cattle there that had been bought and shipped in from various parts of the country. Miller, who people said was losing his memory, pointed out to the doctor and nurse attending him every one of the many brands on the cattle and told them from what section of the west their irons had come.

The old boat story, the one in which old Henry assertedly rode over thousands of acres of land and claimed it for reclamation has many variations. Some say he put the boat on a wagon with four mules hitched to it and drove over it. Others say he had a crew of men, and only hauled the boat from one lake or swamp to another, and rowed across each one, having the crew load and unload the boat without his ever getting out of it. It seems the old man was very punctilious in his relations with the land office.

One day he is said to have cut the cruppers off the harness because the teamster after buckling the straps did not put the ends in the keepers. Another time he cut the halters off the saddle horses because he said, "Halters don't go on saddle horses under the bridle."

Horse Breaking . . .

Through the period of breaking from the time the colt was first caught, hobbled, side lined and ridden, up until the double reins were removed and he was in the bridle a finished horse, there was many a chance of the rider meeting with disaster.

If the horse failed to buck, fall over backwards, or stampede, and went through the green stage of breaking behaving like a little lamb, he never failed to break in two when roped on for the first time.

For some months Paul Apodaca had been breaking a number of colts for the Kern County Land Company at Stockdale, and was having good luck in getting them started. The horses were no meaner nor harder to break than the general run of Land Company horses, all being a Thoroughbred-Standard bred cross. He had not met with any

"Go To the Devil, I Work For Miller and Lux"

serious trouble until the day he went out to rope on a one-gutted sorrell for the first time.

A big "Chihuahua" (longhorn from Terraza's ranch in Mexico) had jumped the fence. Paul was young then and careless of life or limb. He tied his rope hard and fast and made a big loop and took after the big brindle.

Land Company horses were always good fast animals, and he overtook the steer and roped him around the neck and between the forelegs. The scared colt went by the steer and hit the end of the rope. The saddle slid back on his rump with the cinch around his flanks. Paul fell clear of the frantic kicking, bucking horse and lay in the grass to hide from the steer, which went on the prod and charged the horse, who kicked out of the saddle that had slid over its rump and on down its legs.

The big "Chihuahua" hooked the flying saddle as the horse ran off unhurt, then turned and charged Paul. The buckaroo managed to grab its horns as he dodged, and despite the steer's efforts to shake him loose, hung on. When on the point of exhaustion he succeeded in throwing the animal, he got its tail between his legs to prevent it from rising. Holding on to the tail, he got his breath back and had quit panting when Emiliano Cordova, who had sensed something was wrong, came to his rescue.

When Emiliano had put a rope around the steer's hind legs and was holding it, Paul caught his horse, saddled him, and proceeded to finish the job of bringing the steer home.

With a rope around its horns he led it to the corral, where Emiliano caught its hind legs and threw it. When the reatas had been coiled again and the two men were riding back, Paul recounted his experience. Old Emiliano remarked with some impatience that breast collars were no good hanging in a tack room but were made to use on "one-gutted" horses like the sorrell, and if young vaqueros didn't have sense enough to use them it was better to throw them away, an observation which Paul has never forgotten though the old man uttered it over 35 years ago.

Brave Old Men . . .

Almost every ranch carried one or two old men on the payroll. They could hold their own or even better the best young men in the daily work of the range. They were old timers who never lost their nerve, the spirit of youth or the instinct for adventure. They were still afraid of nothing though their hair was white and their weatherbeaten faces bore the stamp of years.

Though they groaned all night in their sleep after a trying day and were the objects of muttered abuse from the boys they kept awake, when morning came they were always ready to accept the challenge of a struggle with man or beast, or to enter into any adventure they might meet to prove that age could not conquer them. Perhaps it was pride of their profession, or the innate urge for strife, that made them the first to undertake a chore that some young men would hesitate to do.

Victor Ortega would put a dollar on each stirrup and ride a bucking horse without losing them. Juan Perez, nearly ninety, could rope a hundred calves before he missed a throw. Tiny, shriveled old Avelino Martinez, tucked a six shooter into his shirt to command the respect he got from all men. Jesus Lopez, sat aching on his horse until all his men had eaten before dismounting to eat himself. Bruno Contreras picked his old body off the ground when bucked off by the white horse and climbed on again for a repeat performance. Juan Gomez mounted on a spoiled horse that had thrown an able-bodied youth twice, and gave his gentle horse to the boy to ride.

Old Cholo Valenzuela would take a green colt, ridden two or three times, teach him to stand, then take two boxes and make a platform to mount the colt from. Jesus Garcia would ride a little roan mare for a week working with buckaroos that changed horses every day and do more work than any of them without hurting the little mare.

David Sesenia would mount a horse with a loose saddle, and reaching down give the latigo a pull until the cinch was tightened as the horse bucked.

Joe Alvarado was a fellow whose shirt was always pressed and clean, and whose boots never lost their shine no matter how dusty the day. He was a man who could bulldog steers from a motorcycle. He was a good roper, rider, and knife-thrower who turned out to be a cattle association detective.

Juan Bravo, a noted early day horsebreaker, once threw a loop across the Calloway canal to rope an animal on the opposite bank. Paul Apodaca, old-time Tejon vaquero and still riding colts around Bakersfield, was mailman there when a small boy and rode in for the mail on a black burro called Chanate (black bird). Typical of his breed, the little animal took good care of himself and once, arriving at the stables in Bakersfield would never leave until late afternoon, in spite of all the efforts of his rider to get him going, he would balk until after he had time to dine leisurely.

Vaquero Hardships . . .

Leonardo Chavarria, eight-one years old, one of the four old men remaining who were vaqueros in their youth. Though miserably paid and housed, if at all, existing on a diet of beans and *boludo* (lump jaw) beef, most often broiled on coals, no other class of men suffered more hardships in following their occupation, without swaying from the path of loyalty and the job in hand than the early vaqueros.

In summer they fought mosquitos, in winter they froze, night herding the cattle or horses, for there was no place to enclose them in that era of open range. There were no fences of any kind.

Leonardo has had in all of his eighty-one years, but one mishap with the law. He got a ticket for double-parking his car one day in 1936 when he went into a bakery to buy a loaf of bread, and paid a one dollar fine.

Empty Saddle . . .

Bill Borgwardt was boss for Miller and Lux. The buckaroos were camped out at Salt Slough one day when a horse bucked with Angel Martinez up to the kitchen steps, and as he reached them, Angel sailed through the door and landed under the table. The horse ran off across the open country, an occurrence dreaded by all cowboys since a saddle represented a number of months' wages. A horse running with an empty saddle for any length of time will slip the blanket which works out from under the saddle. It then turns under a horse's belly, where the average ranch horse will kick it to pieces, or step into the stirrup leathers and tear it apart.

Leonardo was riding a big-footed stocky horse he liked very much. Rafael Cuen, the foreman, said the horse was no good which made Leonardo angry. He was determined to show Cuen his error. The men chased Martinez' horse but couldn't catch him, and finally Leonardo told Antonio Maria Lopez to chase the horse toward him. He then set out in a run, the runaway following him, and as he came up to his horse Leonardo reached out and caught the reins, making Rafael Cuen eat his words.

Chihuahuas . . .

When the railroad came in and fences were built, cattle were shipped into the country from the ranches of Luis Terrazas. The cattle were called Chihuahuas after the Mexican State from which they came. Don Luis could sell trainloads of cattle all of the same color, so vast were his herds. He sat in the Hotel Del Norte, with his booted feet on a chair and received ten dollars for each head that crossed the border in a stream, that ended only after years of strife in that country below the border.

Easy Does It . . .

Domitildo Castro had once been very expert at tailing steers. He had grown old and had long since retired from working cattle and was plowing with a team on the day George Coffee, foreman for the Land Company, and three vaqueros passed with a bunch of cattle. One steer went on the prod and got through a gate. When they attempted to drive him back, he charged each one in turn.

Domitildo watched the performance for awhile, and finally stopped his team. Taking off his coat, he went up to the steer. Using his coat as a cape he got the steer to charge so its rush carried it through the gate and back into the herd once more. Putting on his coat again, the old man resumed his business of plowing.

Contests . . .

There were no rodeos as they are known today, as roping and riding were commonplace, and unless money was bet heavily, horse races had little interest to the spectators.

Leonardo Chavarria remembers when there was a sort of contest which caused some interest. An Indian named Burk Burnett and Montana, a Tejon vaquero, were matched to ride five mules apiece. Each was to rope five horses by the front feet for a purse of fifty dollars, a large sum in those days.

George Coffee, George Tibbitts, and Robertson were the judges. Burk Burnett rode his five mules out of the chute one after the other. Montana, to beat him, leaped on each mule from the ground as it went out the chute gate. Burk Burnett missed one throw out of five in the roping contest. Montana caught four, then threw with his foot to catch the last horse with a perfect *mangana,* winning the prize of fifty dollars.

At one of these impromptu contests, the boys were engaged in a game of *correr al gallo* (literally "run at the rooster"), on the outskirts of the little cow town of Bakersfield. To make it risky, the rooster was buried near the end of a bridge, so the hanging rider would be forced to get back into the saddle immediately after reaching the rooster to avoid hitting the bridge. Most riders would be back in the saddle before reaching the rooster. The bridge would scare them.

Montana was riding a big black horse that belonged to Alex Godey. With one hand grasping the mane, he leaned down. Stretching to reach the bird from the tall horse, his hand slipped and he fell. But he was so active and quick on his feet, he recovered and leaped back on the horse before the animal reached the bridge.

Some little skepticism could be expected when the old men told of pulling a rooster out of the ground when hanging from a running horse's back with the back of the hand dragging along the ground, and the fingers spread to catch the bird around the neck. Most range horses classified as gentle vaquero horses would have kicked the head off the rider who would lean down and hang, making too perfect a target for the horses to ignore. But since all old-timers speak of it as a common pastime, what they say must be true. The horses were picked for this game and were naturally gentle. Some men became very expert at this *correr al gallo* and were naturally combative, because the instant a rider succeeded in pulling the rooster he was immediately assailed by the other riders trying to take the fowl away from him.

The man who held the rooster had the privilege of beating his assailants over the head with it. Many fiestas ended up in a free-for-all in which a good time was had by all.

Baptiste Albitre, a good all-around man, was very good at this game, and trained his horse to run straight and true past the buried *gallo*.

Cavalry Horses . . .

The government bought cavalry horses here in 1880, and Conducci Castro, Leonardo Chavarria, young boys then, with a man from Santa Barbara named Morales, broke 67 head in fourteen days. Morales, a very good rider, being a grown man, took the lead in the tough job of breaking a pale buckskin horse. It threw him, breaking his collar bone and a leg.

Such institutions as compensation and insurance were unknown to the vaqueros. They dug down into their pockets to help pay for the doctoring of their *companero*. Each of the boys gave ten dollars, and the man who had the breaking contract with the government gave thirty dollars to help Morales through his illness.

Cimarones . . .

In 1886 Tom Briggs quit working for Miller and Lux. Luis Johnson became foreman and Leonardo went to work for him. The spring roundup was on. They had started to make a circle toward Dobie Hole when Anderson caught sight of a band of wild mustangs near where Lost Hills now is. Boss Johnson said to let the cattle go and run the mustangs. So they started after them, all of the riders in a group. At certain distances one man would drop out until only one rider was left to run. He followed to Panocha where he turned the mustangs. Then the men who had dropped out took over the chase as they reached each one like a relay. Leonardo took the *cimarones* at Lost Hills, and at 4 o'clock drove them into the corral at Dobie Hole.

All the men were fired from running wild horses, but Leonardo didn't lose anything. There was a horse with a Kern County Land Company brand, running in the lead. He knew the horse and roped him. He later turned the horse, a bay, over to the company foreman, named Hunt, who paid him $20 reward.

Frank Pepper, who came on as foreman for Miller and Lux, hired all the men back, so no jobs were lost.

Skyliners . . .

Orville "Jack" Glen was reminiscing of the early days when there was much of the Andalusian strain in the cattle and horse herds ranging here. Ranchers owned cattle that died of old age in some brush thicket without the owners ever catching more than a fleeting glimpse of them from year to year.

Elusive rawboned old steers were always on the alert, watching and listening for their enemy, the vaquero. When hunted in brushy country they would enter a thicket and stand without making a move or sound until the man hunting them was gone, never knowing his search often brought him within a few feet of the hiding outlaw.

After being surprised and chased a few times these old steers would learn to follow the ridges when traveling and would spot any rider in quest of them. They seemed to sense an enemy when he was yet miles away and would break into a mile-devouring trot, which covered astounding distances in a day. The vaqueros called them *skyliners.*

They were always ready to fight, and immediately went on the prod when overtaken, roped or cornered. They became wilder and more furtive through the years. *Orejanos* and bulls took up with them and when vaqueros hunted *orejanos* (unmarked calves) which had escaped branding, the riders would first find the steers' feeding ground, sure to see a number of the animals. One time Glen roped an outlaw and got a tree between himself and the charging steer. He tied him to this tree and went home to get a saw as the steer had horns that grew straight up, then curved back over his shoulders. The steer fought and butted the tree until he ground loose the rope tide around the base of the horns, and when Jack got back with the saw the old skyliner was gone. Another time he and a man named Arbiso roped the steer and tied it to a pine tree with a hair rope. The pine was on a steep hillside. Jack got off and was pulling his rope off through the hondo when he saw the hair rope parting. He yelled to Arbiso to look out and made a beeline for his horse. When the rope broke, the steer standing on the steep hillside fell backwards and rolled over a couple of times giving the two men time to reach their horses.

Jack finally traded him to Arbiso for a load of hay and the new owner took a thirty-thirty and went out to hunt the old outlaw. When he cornered him against a cliff and dismounted to shoot him the brute charged Arbiso and it took five shots to bring him down a few feet from the rifleman. Hides were cheap in those days but the old steer's hide was so big it brought $10.50.

Being short of meat one winter Jack went out with a rifle to hunt a skyliner just as he would a deer. He came to a little bunch upon a ridge and shot at one. The steer went out of sight and he thought he had missed. He shot at another which also went out of sight. He climbed upon the ridge and found two dead steers lying in the snow.

Jack remembers that there was no grass for their horses to eat so they fed them leaves from the scrub oaks and barley. The horses stayed in good shap through the three or four weeks they roped, and tied wild cattle in those mountains.

Mustangs . . .

Wild mustangs were numerous on the ranges and Jack observed many things about them. The *broncos* were pursued and overtaken by men mounted on horses taken from the same wild band. The mounted ones could outrun the wild ones because they were fed grain and this food gave them much more stamina and endurance than those which subsisted on grass alone. He remembers a friend of his who followed a black stallion for all of fifty miles. Later while driving the *bronco* who had his tail between his legs so exhausted he could not lift it, they came to a gulch. The wild one jumped, missed the far side and broke its neck. The horse his friend was riding was to all appearances still fresh.

Jack says a biting horse will kill a kicking horse, and when a stallion that uses its teeth in a fight meets a horse that depends on its heels the biter will be sure to kill the other. Wild horses when fighting spar for an opening like human fighters or wrestlers, and when one gets a death grip on the neck or withers of its opponent it hangs on until the other is dead.

When a stud sees a rider approach he will start his mares in flight and when they are moving he will turn around, lay his ears back and with his head along the ground, charge up to within twenty or thirty feet of the rider, a magnificent bluff to give the mares time to get away. A wild stud, when cornered, will make a desperate effort to get away. He will weave in and out of the herd until he finds an opening between the riders surrounding him and then make a break for liberty. But the vaqueros would usually shoot him before he made good his escape.

The Tulares . . .

Jack Price who has seen seventy winters come and go, was born in Sacramento, but came to the southern San Joaquin long before the cotton and potato fields had begun to rise out of the vast tulares that covered the land.

These tulares were the refuge of outlaws and fierce wild hogs. A fugitive could hide from an army in the miles of rushes that rose high overhead. But even there men were not safe for the deadly malaria lurked unseen in the reeking swamps to strike down the furtive horse-theives who chanced fever and chills to escape their fellowmen.

There were myriads of wild fowl to rise in loud protest and warn inhabitants that an enemy was near at hand. The wagon road clung to the skirts of the hills as if in fear of being swallowed by the clutching quagmires.

Old Bakersfield . . .

Jack still rides after cattle in country so rough his horses must be kept constantly shod. He doesn't shoe any more so a friend and I went up to his place at Lorraine to do that little chore for him.

After his string had been taken care of, we tarried there to yarn a little. We went into his tack room to admire his silver overlaid Spanish bit and fine bridle reins of braided calf skin, for Jack is a true Californian and adheres religiously to the old Spanish style of equipment and methods of working livestock. He reminisced of the days when he first came to Bakersfield.

He said that one store would have a board walk, the next one would have none. During the rainy season a person could walk dry shod over the boards, then step down to wade through a sea of mud to the next walk. Twenty-horse jerkline teams would come into town, the horses keeping step to the sound of the bells on the leaders. The wheels would spray an arch of mud away from the wagon. Now and then a lady in voluminous skirts would be forced to shrink and cower into a doorway to escape this rain of slime

Catfish Spurs . . .

Catfish abounded in the irrigation ditches and a mess of them was easily caught.

Old Thad Chaney, a rough and ready buckaroo, was riding with Jack one day when they came to the weir at Bellevue ranch. The water had been turned off and a pool near the weir was full of fish. Chad, being very fond of catfish, got two gunny sacks and half filled each with the denizens of the pool. He cut a hole near the mouth of each sack, hung them on the saddle horn and mounted the horse. Catfish have sharp spikes which, after he had ridden aways, worked through the sacking and had the effect of spurs on the horses shoulders. Jack was some distance ahead of Thad when he heard a commotion and looked back to see Thad's horse bucking wildly and the sacks swinging violently. They couldn't come off the horn because Thad was hanging on to it for dear life. Finally a sack hit Thad so hard it knocked him off his horse and he landed on the ground in a sitting position where he could watch the horse running and bucking along the road to Bakersfield stringing catfish all the way into town.

Wild Mustangs On the Ranges

Buckeroo Bosses . . .

Droughts, barbwire or bad luck would sometimes have so serious an effect on the colt crops, that in a year or two, the ranch would be short of saddle stock. One of the other ranches owned by the same company would send horses to supply the mount required for each man's string.

Foremen welcomed these opportunities to get rid of their meanest horses, and it is needless to say that the ones which bucked, fell over backwards or had to be hog-tied to shoe, would be picked out and sent. The milk of human kindness curdles early in most buckeroo bosses. A foreman, no doubt, grinned in fiendish glee thinking how rough a ride the outlaw horse he was cutting out would give some vaquero on some distant ranch. This is how mustangs from the White Horse in Nevada, the Malheur or Harney in Oregon were to be found in the *caballada* (horse herd) on the Buttonwillow ranch in Kern county.

These new mounts were always regarded with suspicion by the buckeroos. They were an unknown quantity, so to speak, to be approached with caution. On the first time up the vaquero would discover whether he had a bucker, a *cinch binder* or one that would *coldjaw* and stampede with him.

The Gringo Kid . . .

The vaqueros on Cholame ranch in the hills of San Luis Obispo county were standing in front of the cook house one hot day when one of their number called attention to a rider approaching from the direction of Bakersfield. The rider was a lanky kid on a little brown horse that dragged its hind feet. The dust of a long dry road covered both horse and rider. The boy rode an old slick-fork saddle that had twenty six inch tapaderas. He seemed to be holding them up with his feet to keep them from dragging the ground. But his Spanish bit was silver-mounted and the metal had been polished at some recent date. A well-braided service-worn riata hung in neat folds from the strap on the fork. His equipment indicated he had been schooled by paisanos and had adopted their methods.

When he rode up the boss asked him to dismount and stay for dinner. He pointed out a feed rack where he could tie his horse. The boy dismounted with alacrity and tied the little horse to the manger where the animal fell to eating eagerly in great greedy mouthfuls as if he had been on short rations for many days.

After he had washed up the kid proved to be one of those blond, blistered, red-faced, pale-blue eyed types the appellation gringo most readily applies to. He spoke Spanish and this set him apart from others of his kind. The vaqueros sensed in him a mutual sympathy and understanding for them and their ways. The boy evidently had been brought up by Spanish people. After dinner the men trooped out to look over the little horse.

The beast had a *Roman nose, sheep* head, lean flanks and round drooping quarters. Jaded as he was he snorted at these strangers and rolled the whites of his eyes and looked sideways. Nervous by instinct, he distrusted all man-kind and that buckeroo boss in particular. He stood gingerly on tender, footsore, shaggy legs.

After studying the animal for awhile the boss asked the stranger if he wouldn't trade his tired mount for a nice fresh, fat horse. He said further that he had been looking for a pony for his kids to ride and he thought this little brown would *fill the bill.* The boy said he would like a nice, fresh horse but he would have to trade *straight across* (give no money to boot) since he was broke and looking for work.

In the end the boss traded him a big sorrel that had *rollers in his nose* (snorted violently).

The boy had to blindfold and hobble the sorrel before he could get him saddled. After he had led him around a few circles he was ready to mount. He pulled the blinds down over the horse's eyes, took a hitch at his pants and pulled his hat down firmly on his head.

One of the men said, "Kid, I had better tell you that horse bucks like hell, so look out for him."

The kid mounted, settled himself in the saddle and raised the blinds. The horse squatted for a moment, then *blew his cork* and began to *unwind.* He had bucked a lot of men off and had been turned out.

But the boy rode him and whipped him into a run. When he passed the group of men he checked the horse long enough to call back to the foreman and say, "You may think this horse is bad but the one I left is bad too. You'll find out." Then he rode away in the direction of Paso Robles.

The little brown horse was never ridden by the bosses' children. In fact he was ridden by only one man, and he only rode him once. He remained in the *caballada* fat and *sassy,* a thorn in the *mayordomo's* side to remind him he had been outsmarted by a kid not dry behind the ears. He often wondered who the traitorous *paisano* was who had taught this gringo so well that he could beat a *majordomo de vaqueros* in a horse deal.

Months later a rider from Miller and Lux passed by the Cholame and recognized the little horse as one of the condemned bunch that had been at Buttonwillow. The boy had bought him for two dollars and when he had mounted him the men had hazed the horse in the direction of Paso Robles and left the boy to shift for himself on the long road to the coast.

The years have been kind to the once lanky kid. He is fat, bald and old now and runs a cafe in a little town over on the coast. Whenever a man in boots and cowboy hat happens in he will drop whatever he is doing to ask the stranger where he is from, and ends up by telling of the experiences of his youth, tales which the listener, no doubt, seldom believes.

Changed Times . . .

The advent of the barb wire fence brought to an end the era of free grass and open range. Cattle formerly worth only the price of the hide which covered them became more valuable, and consequently better bred. But no amount of breeding could produce a better stock horse than the Spanish mustang.

A special providence placed the hardy, tireless little animal ready to hand for the vaquero to ride through the period when great herds and cattle empires flourished. Horses, then let us say, were not better bred, though the signs of their having Thoroughbred, Standardbred, and Morgan blood in their veins had become evident in the bands of mares and colts. The rough and ready "break him or break his neck" methods employed by the oldtimer in handling livestock began to be frowned upon and were discarded as much as possible.

Though this change in conditions benefitted the stockman, it deprived the young vaquero of the practice which makes for perfection in a highly skilled trade. The youngster was forced to sneak out on moonlit night to rope cattle and to find some dry canal or spot out of the boss' sight to get his experience in riding bucking horses if he was to approach anywhere near the skill of the old-time open range vaquero who preceded him. The new buckaroo got his practice whenever he could. Consequently, there were many little rodeos and contests pulled off when the boss was away and which he never knew about.

There was always some colt humping his back and trying to buck. As long as the boss was around his rider would patiently hold his head up and double him to keep him from breaking in two. But when the *majordomo* was absent the vaquero showed no such forbearance. He would take the first opportunity to spur the colt a little to see if it could really buck. Young buckeroos are a curious lot and are always trying to find out if they can stay on a bucking horse, since all aspire to be *ginetes* (bucking horse riders).

Envidias . . .

It is thirty long years since Albert Blanco and I rode the range for Warren Stockton. A tale he told me then has stuck in my memory for lo these many winters. It is a story of *envidias* (envies) among the vaqueros, and happened a long time ago when Don Jose Blanco was manager of the great Cuyama Rancho.

In those days the Cuyama ran many head of cattle and bred large *manadas* (bands) of mares and colts. Don Jose employed three *amansadores* (tamers). They were Juan Olivera, Luis Zamora, and Delfino Villa. Luis and Delfino roughed out the *broncos* and Juan finished them. The boss considered Juan the best reinsman of the three. This caused much ill feeling, the two resenting the boss giving Juan the best and easiest part in the rough, dangerous work.

Juan was a soft spoken, retiring sort of man. He had little to say and not much to do with the other men. He always wore a necktie when working and he had that knack of keeping his shirt clean in spite of the dust, dirt, sweat and sometimes blood of the cattle corral. His boots and hat were expensive and of finest quality.

One day when Mr. Blanco had gone to spend a few days in town the boys dug up a gallon or two of wine. After it had been passed around a few times the bad feeling among the men burst into argument and the result of it was that Luis and Delfino challenged Juan to a contest that would determine who of the three was the best rider. There were three unbroken five-year-old stallions in service on the ranch at the time. Luis and Delfino proposed that each ride a stallion and the man who succeeded in taking the halter off his horse while it was bucking was to be considered the best *ginete*.

The two men offered to bet money but Juan said he was no gambling man, but if they insisted he would ride against them for a twenty dollar Stetson as the prize, the two losers to pay for the hat.

Luis chose a stallion and rode first. Luis was a light complexioned gray-eyed frail looking little man, a fine rider and entirely without fear and a good all-around vaquero. He rode the horse in fine style until he attempted to unsnap the halter. He lost his balance and was bucked off. The wine had not helped him any.

Delfino who rode next also fell a victim to the *vino* and *pulled leather* to stay on the horse.

Juan, one of the best balance riders of a day when most men rode in slick fork saddles and used their spurs to clinch the bars of the cinch to keep themselves on the hurricane-deck when a horse was bucking, rode his horse without these aids. He removed th halter while the horse was still bucking and the two men who lost the contest tossed him a $10 bill apiece, and Juan remained the best *ginete* on Cuyama.

Of late years Juan has not been seen on the ranches in the San Joaquin valley. But every now and then one hears of an old white-haired man who breaks colts around Santa Barbara. People say he can do wonders with a horse despite his age. The old man is Juan Olivera. Albert says it was not until many years after Don Jose Blanco had retired from the management of the Cuyama that he told him of the riding contest among the vaqueros *envidiosos* who paid in cash for their envy.

The Little Frenchman . . .

Johnnie Puget, the little Frenchman, got used to sleeping on the hard ground while working on the Miller and Lux chuck wagon. On his first night at home after his initial hitch as a buckeroo, he found the bed his mother made for him too soft to sleep in. After lying awake a long time he spread the blankets on the floor and laid down on them to sleep soundly for the rest of the night.

Johnnie learned to speak Spanish as soon as he could walk, but did not learn to eat chili until he went to work riding the range. The vaqueros vied with each other in eating the hottest peppers and the kid, determined to be in all things like them, ate them too. The first morning at home he got up at 3:30 o'clock, the usual rising time on the ranch and made a fire in the kitchen stove. He was frying chilis to make a vaquero omelet, when his mother, awakened by the smarting, peppery fumes, got up to protest loud and long this desecration of her kitchen.

Being a good cook and proud of her ability she could only shudder at the result of her son's culinary efforts.

Johnnie was telling of the time when he was on the *chain gang* (wagon crew). They had made the long drive with cattle to the New Columbia up near Firebaugh, and had circled back with another herd to Carneros Camp far out on the west side of the valley.

The grub had run low and the men had been eating, as Johnnie put it, "Fried beans for breakfast, fried beans for dinner, fried beans for supper." One morning the last batch of the seemingly endless *frijoles fritos* was sizzling on the fire. Santos, the *cocinero,* was mashing them with a big spoon when the brown S wrench horse, Willie Araujo aboard, bucked through the camp and scattered the fire, pots, pans and dishes leaving the crew without even beans for breakfast.

The brown S wrench horse had come from Nevada, was rolling fat and plenty spooky. Whenever a buckeroo went into the corral to catch a horse, the brown would back into a corner, snort and show the whites of his eyes.

All the men expected him to buck with whoever rode him. Old Jake, the boss, had given him to Willie a few days before, and because the kid hadn't saddled him immediately the other riders had kidded him about it, hinting slyly he might be a little afraid of the brute.

To put an end to the boys' making dire predictions about what would happen when he did ride him, Willie had saddled the horse that morning and got on.

Whether he spilled his rider, Johnnie forgot to mention. This incident and the brown horse have stayed in his memory through the years, largely because he *spilled* the beans.

The Arabic Spur . . .

Now and then some vaqueros riding the range on one of the old Spanish ranches, or along some old road once used by early travelers, will find half-buried in the sand a great roweled spur, lost a century or more by some passing horseman. The band and shank is worked into lacy, intricate design with the crescent of Islam outlined distinctly in the Arabesque ornamentation. Some Andalusian artisan with the memory of home still fresh in his mind had worked the iron into the delicate patterns he had been so familiar with in his native land.

This type of decoration was used before silver had become plentiful enough to be used as an ornament by the common man long after the

Spaniard had been freed of the disease that only gold or silver could cure.

The designs on silver inlays in modern bits and spurs are modifications of those found in the fine ironwork of Arab lands, and the star and crescent appear many times in the work of modern craftsmen who perhaps have no idea or inkling of where it originated.

Thus the early Iberian, having no houses to decorate and living in the camps, decorated his bits and spurs with the delicate tracery of Islamic art.

The ancient Moorish spur with its over-all length of ten inches, two inch wrought iron band and six inch rowells had been well-worn long before it had been lost or left in the San Emidio Hills. But the fine filigree workmanship of the Moslem crescents was still discernable in all its details despite the rust of years. Its delicately tapered rowells come to a fine point and with its triple rowell rocks it still had the appearance of being a brutal and indispensable item in the equipment of some far-ranging horseman of this country's early history.

Pomusino gave me the spur some years back and told me it had been found in the region back of Mil Potreras where according to local story-telling, the Lost Padre mine is supposed to be located. We had both laughed when he told me that because we had each of us known padres in our time, and padres are men who do not lose mines or other things.

So I took it to Felix Guitron who has a large collection of such things and is well informed on the lore of Spanish bits and spurs. He can talk for days on that subject and never say the same thing twice. We placed it among the three or four hundred articles in his collection and he set its date as late eighteenth century. Our conversation turned to the art of using the spur, *a la gineta,* in a straight line from the cinch to the flank. The use of the spur was the supreme art of the Moor, for his life depended on it as much as on the curbs of his bit. In battle those old horsemen relied on the spur to help lift a horse away from the thrust of a lance or sword as the modern day *torero a la gineta* lifts his horse away from the horns of a charging bull. A horse that would not respond instantly to the signal of the steel was unfit for war and discarded as worthless.

So we sat there in his tack room among the treasures of his *aficcion* and indulged in conjectures regarding the rusty old relic. We wondered how it had come to be there in the mountains that rim the great valley far from any Spanish settlement. Was it discarded or lost by some Spaniard who had been forced to walk? Did his horse come to the end of its strength and die there on the bleak mountain, was it killed, or did it wander off into the valley to father the countless wild mustangs that were to be found there in later years? Did some Indian get possession of this spur in trade and bring it back to these mountains for his people to marvel at who little knew that this was the implement with which the Spaniard impelled the strange charging beasts upon them?

Pomusino, who had much Indian blood in his veins suggested it had been spoil from some raid on one of the missions. Here was a memento of a people who went to Spain, who ruled and disappeared and set the pattern of conquest for their descendents in America, who also ruled and disappeared in like manner, the only monuments to their brief stay crumbling adobe walls and old relics such as this rusty moorish spur.

Cow Horses . . .

The long gone Californian like his prototype, the present day *charro,* kept his *caballo de rienda* to show his skill as a reinsman and did not confuse him with the vaquero horse which is a work animal like a sheep dog. If left alone he will follow a cow and remain with it until it is parted out from the rest of the herd.

He requires no guiding from the rider and works better on a loose rein and if left to turn at his own will. This instinct for working cattle is found in all vaquero horses in more or less degree. Californians call it *amor al ganado* (literally, love for cattle). Years ago when a majordomo like J. J. Lopez, Rafael Cuen, or Chico Urrea or any of the other bosses were giving a new rider his string of horses they would class each one in the extent of his love for the work.

But the senior don of the old time grants made a distinction between the horses and often remarked that the horse some vaquero was riding knew more about cattle than his rider and it would be less shameful to the horse if the man were to get off and let the animal work without the added burden.

Bits . . .

The general supposition that the spade was the only bit used to ride reined horses is a vast joke to the men who knew the last of the old time riders. While the *barbareno* or *freno del pais,* as it was called, was common there were many straight-shanked half breeds in use. The ring bit called the *chileno,* from the Chilians, who came here in the gold rush period, was found much in use. The bit used by most men who rode on the ranches was one fashioned after the old Salamancan, which was identical with the old U.S. Army bit, only smaller in proportions and usually more refined and silver mounted.

It is amusing to hear some one say that the curb bit is far less cruel than the Spanish. The person making this remark does not know that the Englishman got his bit from the same land he got his horse. The statement that certain methods have been developed here is also quite untrue. The spade was old in Africa centuries before Father Kino ever set his eyes toward California. One can look at pictures of modern day Ethiopians mounted on mules wearing ring bits identical with the *Chileno.*

Every now and then some *charro* like Jose De La Torre will come up here out of Old Mexico and throw a hundred-foot riata he learned to handle in Jalisco where the long ones are used. Along the coast of Queretaro, Nayarit, and Sinaloa the men who ride after cattle have been tying rawhide riatas hard and fast to the horn for some two hundred years. These men use the *filete de Amossoc,* which is the present day grazer curb bit and originally was brought from Spain.

Felix Guitron of Indio is a member of the Charro association and a disciple of the Marquis of Guadalupe, dean of them all. He owns 300 originals or copies of Morroccan, Spanish, North and South American bits and numerous items of equipment, much of it historical. He has three well-trained horses, two of which have been used to *torear a la gineta* in the bull ring, where a horse must rein or die by the bull's horns. Mr. Guitron used a *filete* (curb) to bit up his horses and is now in the process of adopting them to the ring bit. His reinsman, an 80-year-old expert says it is the only bit to finish a horse.

The King's soldier has gone from the soil of California and the art of reining horses passed away with him. All that remain is a muddled legend of his artistry. We who knew the last of them are happy to know that literacy was not one of his accomplishments and that he cannot read what is written regarding his horsemanship. Because if he could he might turn over in his grave.

Death of Spanish California . . .

The sun had set for the Iberian in California. His numbers had been

Even the Devil Appreciates Kindness

too few and he was too far away from the motherland. He had lost his birthright and his children spoke to him in the Saxon's tongue, heedless of wounds that would not heal. Clutching at the straws of broken pride the old people clug to the tie of language in a brave effort to preserve the lore of a race that was passing beyond recall. The old ones told each other over and over again the simple tales, for they were simple folk. Often a remnant of an old legend in fading Spanish would cling to the memory of a lad who in later years would see through a mist of unfamiliar words the soul of a people with eyes he had not possessed as a child. Memories that left a vague regret and longing for something he had lost and would never find again.

Juanito and the Devil . . .

There is the story of Juanito to prove that even the devil himself will respond to kindness and be grateful. It was often told to guide some youth along the path of rectitude and compassion for weaker beings.

Junaito often wondered why no one ever prayed or burned a *velita* (candle) to *El Diablo* and only spoke evil of him. One day he was near the church when he was struck with pity for the outcast and said, *Pobrecito,* no one cares or burns a candle for you, but I will."

Being very poor, he had no money to pay for a taper. So he went into the church and scraped up all the wax that had dropped from the shrines, made a candle and lit it. Satisfied he had done a good deed he left the church. Years passed, he grew to manhood and married a woman as poor as himself continuing to live in poverty.

One day, having no food to give his family, he set out along the highway to see what he could find. He had traveled some distance when he met a man dressed entirely in black wearing a great cape and riding a fiery black horse. The horseman drew rein and asked Juanito where he was going with such a sad look on countenance. The young man answered that his family was without food and he was in search of something for them to eat. The man in the cape told him to follow and led him over the hills and finally drew up at a cave. They entered and found a mule ready saddled with *alforjas,* (saddle bags) hanging from the saddle tree.

The horseman put Juanito to work loading the mule with gold which lay in heaps in the cave. When the packs were loaded the rider said, "In the many ages I have been a fallen angel, no one has ever had pity on me. You are the first one to have done a kind deed for me. Take this gold and when it is spent come back for more." Kindness is never wasted in handling animals or devils! So thought the poor paisano.

Sonorans Return . . .

Gold had been discovered in California and the Sonorenos were coming over the bóne-strewn highway so aptly called *Road of the Devil.* They were seeking riches or a new life in this golden land, so different from the barren strip of red plateau and thirsty verdureless stretches of burned earth so little worth the owning that they had left behind, never dreaming at the time that regardless of the fortune or fate they met in this land they would be eager to return to their homeland to end their days. It was a common saying that when one got old it was time to return to Sonora to die, as most of them did.

Two Rascals . . .

There were various types of honest men and rascals in quest of the wealth hidden in the Sierra Nevada. One often-told tale of those times is the one of two men who being destitute, used their wits to gather in a few *centavitos* (pennies). A number of families had camped on the outskirts of Bakersfield under some trees. The two rascals arrived there and found an abandoned table under a cottonwood and hit upon a scheme. The leader of the two sent his partner into town to buy some candles. On his return with them, the leader instructed him to lie down on the table. When his partner was stretched out the leader lit the candles and assumed an attitude of deep sorrow. Soon his neighbors came over to inquire about the dead man and sit with the body. The leader told them his partner had died of hunger. It is customary to serve food to the mourners and when the visitors saw that none was forthcoming they left and brought back food and a small donation to bury the dead man. When 12 o'clock came, all the food and money having been collected the corpse sat up scaring everybody away. The two rascals gathered up the articles and went on their way rejoicing.

Smith and Packwood . . .

The cattle industry has been a way of life to the family of Hugh Smith of Granite Station for more than 100 years. It was in 1850 when his grandfather, Jake Smith, and his granduncle, Elishu Packwood, started from Missouri with a small herd of a hundred head of red Durham cattle. They traveled with a wagon train across the plains, penning the cattle every night in a corral made of the wagons joined in a circle as the herds of buffalo would scatter the cattle and it would take a number of days to gather them again.

It took six months and eight days to arrive at Grass Valley in California, and in 1853 Packwood and Smith moved to Porterville and built the first house on Tule River. They ran their cattle on land where the city of Porterville now stands. Their vaquero boss, *Romauldo,* was killed in the Indian uprising which required the soldiers of Fort Tejon to come on a fast march to quell it.

Packwood planted barley in 1864. The following year was so dry the barley never sprouted, and the year 1866 being wet, the barley that had lain in the ground over a year sprouted and he got a big crop. Mr. Smith can indeed say he is a pioneer cattleman.

The Horse Breaker . . .

Horse breakers are kin to the gladiator. There is a steady alert look about them which is characteristic of the men who fight broncos and stake life and limb on their skill and quick thinking to earn their living.

Old Sam Howe's features are grim. His body is erect, flat backed and his shoulders have the stoop long years in the saddle give the horseman. He has bowed legs and a stiff walk. His limbs have been broken in battles with horses. Put him on a horse and he becomes graceful, skillfully handling his mount with smooth expert hands.

His life has been a series of different horses. He is old in experience and wise in the ways of the horse. He can judge a horse and read his character like a book, and classify him from his general appearance, and know at a glance how hard or how easy it will be to break him.

When he looks over a corral full of colts he knows there are some there he will dread to climb aboard for he is old now and his bones ache for days when a horse bucks with him. But he will go ahead about his business of breaking steadily, quietly, patiently, using every trick he knows to make the animal gentle and submissive.

He knows a horse need not buck when it is being broken, and if he never bucks so much the better. He knows, too that a horse learns how to buck and will try until he throws his rider. He knows that one can never ride the buck out of a horse, though he can ride one when he does buck. For a horse under certain conditions will do the same thing over again that he has the habit of doing.

Then, too, when a man is afraid of a horse he scares the horse and the animal will fight the thing he fears, and a man will get hurt if he is afraid.

This is Sam Howe who has had many rough rides on bucking horses since he started riding for Miller and Lux in 1908, and has spent forty-two years as a buckaroo, rodeo bronc rider, bulldogger and roper. He has been a packer and cattle buyer and is still breaking horses and doing a good job of it. Born in Bakersfield, he was brought up here and once escaped death in the clutching, quick sand of Kern river by grabbing the tail of a dead horse as it floated by and hung on until the force of the current pushed the body by, and pulled him out of the sand to a clump of willows where his companion, Salvador Rodriguez, pulled him out of the river. The *Paisanos* called it *Rio Bravo* (Fierce River) for it took its toll of lives even in those early days.

Caballo Chino . . .

The *Caballo chino* (cury-haired horse) figures quite often in the tales told by the old-timers who say the *chinos* were *muy guapos* (very tough) and not uncommon in the bands of mustangs found here in early days. Al Mora told me this story of Pajaro, the black *chino* horse, who traveled from San Carpojo to Lebec in a night and half a day.

Juan owned the best-reined horse in his county, back in the days when life was cheap in California and the taking of it frequent and commonplace. Juan, an innocent man, was pursued by a posse eager to catch and hang him. He was a man who never lost hope or his nerve, and when they closed in on him he made a feint by dashing to one side of the circle of riders then, wheeling, he dashed back. He made several feints before he found an opening; then rode between two members of the posse and got away. But the posse hung at his heels and dogged him across the Carriso plans, over the Temblores, and turned back only after they were well into the San Emido mountains. The black *chino* galloped on. Every now and then Juan dismounted to tighten the cinch but never drew rein until he trotted into the yard at the house of Ramon Dominguez, 110 miles from his starting point.

Malicious Mares . . .

We were weaning that year's crop of colts on Rosedale Ranch and had put a band of mares and colts into the corrals. We sat on our horses outside and watched the brutes mill 'round and 'round. Sometimes horses act a good deal like human beings. Mares with colts are sometimes extremely cranky. The three old geldings among them stood quietly while the hysterical females with their lean ewe necks and hollow quarters went on crowding and jostling, getting in each other's way in a causeless panic.

Though their tongues were silent their flattened ears and rolling eyes spoke all their malice. They snapped at their colts, spinning them from one corner of the corral to the other, greeting them with their teeth wherever they might stray. Now and then some mare would deal the corral wall a kick vigorous enough to break a leg.

Some of the colts sought safety in a side pen, but there stood another old mare with her head down. There was a clear space all about her and there she stood, motionless, ears laid back, ready to charge anything that ventured within reach of her teeth—a living picture of narrow-minded spitefulness. One of the men watching her from outside the corral remarked that a mare will give more trouble than 10 geldings.

Gonzales' Murder . . .

Thirty-five years back the ruined walls of an adobe were still standing in section 10 near Connor's station by Lakeside Ranch. Rawhide thongs were still tied to some of the rafters. It had been the home of old man Gonzales who ran cattle in the vicinity and was credited with having brought the first *mesquite* into this country. He built a corral of willow poles which sprouted and made a circle of living trees, still there at that time.

Gonzales sold a bunch of cattle in town for a large amount of cash and returned home with the money. He was not seen about afterwards and C. L. Connors and two other men went to his house to find the old man's body spread-eagled to the rafters, tied with rawhide thongs. There was evidence of torture on the body.

Whether he told where his money was before he died or whether he kept his secret and disappointed his slayers was never known. Connors and his companions cut him down and buried him. *Paisanos* were careful to avoid the place afterwards and no one would live there again.

Bruno Contreras . . .

Bruno Contreras had the reputation years ago, among the *paisanos* here, of having been a vaquero for Pio Pico, last Mexican governor of California.

We were moving some horses from town to Greenhorn one spring in the early twenties and stopped to rest at Mun's ranch on Poso creek. There was one man on an ancient white horse roping calves in a corral. One man and a young boy were tailing and branding. They were having a tough time, so my partner and I entered the corral to help. The roper was very old and almost blind for I noticed he would cock his head to listen for the sound of the calves' movements, then throw his loop. His arms seemed affected with palsy, his movements were so very slow, but when he took his dallies there would be a calf caught by two hind feet at the end of his rope.

I remarked about him later to some people who knew him well and they said the old horse bucked Bruno off once in a while even at that age but the old man would ride no other.

The Goat Lover . . .

Another early day character spoken of by old timers was an old man who loved goats and hated sheep. To avoid killing his own animals and yet at the same time have meat, he would hide in a culvert whenever a band of sheep came down the road. When the woolies passed by he would catch a nice fat one, cut its throat and wait until the band was out of sight before leaving the culvert with his supply of fresh mutton.

A Little Leather . . .

We were driving the cows away from Deep Wells where we had left their calves to be weaned. One old cow, remembering her calf, would keep breaking back causing no end of trouble. Don Francisco told one of the younger boys to double his reata and bring it down over her back a few times which the boy did, and the cow entered the herd and went along without giving us any more trouble.

Don Francisco chuckled, "A little leather, laid on now and then fixes things wonderfully." He went on to tell me the story of the meek man, his shrewish wife and the stubborn mule.

"The husband was hard at work in the corral—feeling thirsty, he called to his wife to bring him a drink of water. She not only would not bring him water, but began a severe scolding; and to escape her tongue and find peace, he walked off along the *camino* to a grove of trees where he sat down facing the road.

"After a time an *arriero* (packer) with his train of mules, drove up.

There was a log across the road and each mule passed over in good order until the last one came up. This animal snorted at the log and turned back down the road. The arriero drove it back and made two or three attempts to drive it over, but the mule proved too obstinate.

"Finally the arriero lost his patience and untied the *tapa ojos* (blinds) he used to cover the mules' eyes with when he loaded them, and carried attached to the saddle. He gave the mule a good tanning and when he was through, it was glad to jump over the obstruction and go on with the others. The husband watched the whole performance and approaching the *arriero* he said, "Senor, do me a great favor and sell me that leather you used on the mule. It seems to have a special virtue." The packer answered, "I can't sell it because without it, I could never hope to handle my mules."

But the husband persisted saying, "If that strap can make twenty mules behave, perhaps with it I might tame one woman."

The *arriero* was a married man and in the end he sold the *tapa ojos* to the husband who went home with his purchase and back to work in the corral. In a little while he called for water and receiving none, he went into the house, put a bench in the middle of the room, then catching hold of his wife he unlimbered the leather and began using it exactly as the *arriero* had done.

At each blow he would say, "Jump the log you stubborn beast," After a few hearty whacks his wife jumped the bench at command. Laying the leather where it would be ready to hand in the future, he went out to work again. Eager to test the effectiveness of the leather in inducing obedience he called for water. His wife came running out with the gourd and when he had drank, she said, "Dearest, how many times do you want me to jump the bench?"

The cattle had quit bawling and had strung out in a long thin line toward Temblor by the time the old man had finished the tale he had told to illustrate the effectiveness of the doubled reata on recalcitrant beings.

Switching Colts . . .

The old timers would often remember and recount the good qualities and virtues of the horses they rode in their youth which according to their tales were models of perfection. But very few of the old men had anything but evil to say of the long-eared mule. Don Francisco, however, had a story wherein that much maligned animal was a hero and saved him from being captured and perhaps hurt by the friends of a boy with whom he had fought on a visit to town.

The big ranches, where many mules and horses were bred, followed the practice that is called *ahijar* (switching colts). An operation requiring infinite patience and much labor which consists of forcing or inducing mares to accept and mother the offspring of another female, and of a different species. The bands of burros on those ranches were made up of selected rangy, big boned animals. When crossed with the horse, their offspring, the hinny, was first class saddle stock with smaller ears, and was faster and more spirited than the mule.

This hybrid neighted and had more the characteristics of the horse its sire than of its dam the burro.

One day while riding through a pasture he came upon an especially fine, newborn burro foal, and from this chance encounter resulted a chain of events which makes this story. Don Francisco dismounted to throw sticks ats its feet to make it jump and show its action. After watching it move about he decided this fine fellow was fit for the stud.

He went back to the ranch where he was informed Gaspar Alvarado's mare had also foaled a filly colt that day. Two men were sent out and brought in the jennet and her colt. They put them in a corral next to the Alvarado mare. The men then put blindfolds on the mare and the jennet. They then switched colts, putting the burro foal in the mare's corral and the filly in the jennet's. Then one man milked the mare and rubbed the milk over the baby burro. The blindfolded mare let the alien suck without kicking. The burro was then milked and the filly rubbed with the milk.

But while the jennet had no aversion to the filly that little animal shrunk from adopting such an outlandish mother and it took some effort to get it to nurse. The burro foal was sucking the first day but the filly took three or four days before it overcame its fastidiousness, and accepted the strange nurse.

When the men were sure the mothers had adopted the colts, the blindfolds were removed and henceforth the burro would think himself a horse and run with the horses, and the filly would think herself a burro and never leave the long-eared animals.

In time the burro was turned out with a band of mares and bred many good saddle animals, and the filly with which he had traded mothers grew up to foal a sorrel mule colt, clean-limbed and active that turned out to be a mean kicking wildcat, trigger-reined and very fast.

Breaking the Colt . . .

Around the ranch were many tall trees. Chains were tied to the branches, and hung down for a horse to be tied to. This method is superior to the hitch rack as a horse can't get a hard enough pull from this angle to break a rope, nor can he rub his bridle or hook his reins and break them as often happens when hitched to a post.

The sorrel mule was hard to break. If Don Farncisco blindfolded her to mount, she would listen and strike at the sound of his moving or whirl and kick at him. He would climb the tree, slide down the chain and onto her back. Once mounted, other than trying to bite his legs, the mule would be docile and never buck or jump.

The Payoff . . .

One day he rode into town and left the mule tied in a stall next to the loft ladder which he climbed up to dismount leaving the mule saddled. He went downtown where he had a fight getting the best of an opponent who got his friends together and surrounded Don Francisco in the stable.

The fugitive climbed up into the loft and down the ladder onto the mule's back and waited. The bridle he left hanging on the saddle horn. His enemies came in the back and front doors and moved in on the mule. When the rider thought they were close enough, he cut the rope and spurred the mule which spun around and as the men moved in to catch it, the animal lashed out at them, first with its teeth then with its heels and instead of closing in the attackers turned and fled. Don Francisco rode out through the door at a run and while he was shot at, escaped injury and got up into the mountains where he went into hiding not for fear of the men in town, but for fear of the whipping his father was sure to give him if he were to catch him.

Don Francisco said it was a month before his mother could prevail upon his father to forgive him for being *escandaloso* (scandalous) and he could go home and eat hot food and sleep in a bed.

A Father's Blessing . . .

It has been many years since Don Francisco de la Para *arrendador de nota* (reinsman of note) told me this story:

The Reluctant Curandero

Don Francisco had reached the age of 16 years before he dared beard his father, Don Andreas, and ask for permission to go out into the world and make his own way. Parents were stern and strict in those days and children respectful. Boys of good family were brought up in an austere and Spartan fashion.

He told me of going into Don Andres' room and standing patiently until his father chose to speak. He was a youth of the *gente de casta* (people of caste) brought up in its rigid rule of conduct and tradition of honor and family, facing a sire who had never let him forget the code of a people who bear swords and own old rusty lances. Don Andres asked why he wanted to go out and work when he had everything he needed at home. The boy answered that what he had was not his, but belonged to his father and he would like to earn money to give to his mother and to buy presents for his sisters; whereupon his father said his mother was provided for and his sisters needed no presents, but if he wished he could leave in the morning and for the present, go to bed.

Don Francisco remembered every detail of his first odyssey, though it happened 65 years before he told it to me. It was 3 o'clock when his father awoke him and told him his breakfast was ready. His mother had taken it upon herself to serve him his coffee. She had a very serious face as she poured out a cup full. Don Andres had a pigeon gray horse he kept for special occasions. He saddled him while the son did likewise to his own, a *grullo* with a black line down his back and a stripe across the shoulder marked like a zebra and equally as tough. His father, without a word, set the pigeon gray at a running walk which brought the *grullo* up to a slapping trot to stay at his heels through the long morning. Don Andres had said not a word as yet.

When they came to the Camino Real at the boundaries of the rancho, his father stopped his horse and pointing, said, "There is your road, Go!" Don Francisco did not move. Presently his father said, "Bueno, why do you not go?" The boy answered, "I am waiting for your blessing, Padre." Don Andres spun his horse around and set off for home, saying "There will be no blessing." The son followed at his heels and after some distance the father stopped and asked, "Why do you follow?" "I am following you for your blessing, Padre." There in the dusty road the father removed his hat and gave the boy his blessing, and Don Francisco rode away.

Late the next day he came to a large house on a great ranch with quarters for many people. Don Gregorio, the *patron,* knew the father and hired the boy to oversee the cattle work. He was told to report back in three days.

He rode back home to tell of his good fortune and the next morning Don Andres accompanied him back to the ranch where they met the patron and manager. Don Andres greeted the ranch owner and said, "Don Gregario, my son tells me you have seen fit to hire him and I have come to deliver him to you, but I wish to warn you the boy knows nothing, *absolutamente,* and is a *sinverguenza* (shameless one). Henceforth you shall have complete control and shall decide for him as his father, and he must answer to you and not to me."

With that Don Andres mounted his horse and rode away. Don Francisco followed. After a while the elder man turned and said, "Did you not hear? I have left you with Don Gregario, I am no longer your father." The boy answered, "I am waiting for your blessing, father." The elder blessed him and he rode back wondering whether the relentless old man loved him, but determined by his conduct to bring no dishonor or ridicule on his father or his house, for *casta* meant much in those days.

He was sent out to oversee the cattle but his real job was to prevent cattle killing. In those times a hide was worth 12 *reales* and big steers would be killed and skinned, the meat left to rot unused. He stopped the killing of cattle and in the years he spent on the ranch, all men feared or respected him because of the horses he rode.

No matter how armed a man, who knows what a reinsman can do on a good horse, will not dare contend against a rider that rode such horses as Don Francisco mounted. A squeeze of the knees, a whirl and a flip of the reata and a man's arms would be pinioned and his body dragging him to death before he could draw knife or gun.

Since the first stockmen to come to the New World left Salamanca in old Spain, the birthplace of the American cattleman, the reata has developed into a useful and indispensable tool and a potent weapon in the hands of an expert, much superior to gun or knife.

Caballeros . . .

The refrain of some old sad song, a crumbling wall, a Spanish name, a broken sword with an inscription, the words long since rusted out, remain to remind us of the caballeros who *carocoled* and sang in other days. Though the last echoes of their serenades died away a century ago, their spirts still haunt the adobe ruins left in California. Old *piasanos* say that on certain nights one can hear the jingle of rowells on belled spurs and the whirr of the cricket on Spanish bits over the hoof-beats of phantom mustangs.

They say it leaves a strange nostalgia for a time and place they never knew. Perhaps it is the folk tales they heard as children that brings this feeling, for as they often say:

"What merit lies in high sounding mottoes on Toledan blades and peerless horses when each son of a somebody (hijodalgo), one after another, was swindled out of and lost the lands left in his trust by a Christian king."

In shame one after another the descendants of the king's soldiers went back across the Colorado as their ancestors the Moors had gone back across the strait into Africa. Like them they stopped to take a last tearful look back at their native land sing La Golondrina. Then turned their horses' heads southward and rode on never to return.

In Sonora the woman told the story to the children, and taught them to sing the words so the story and song will not be forgotten. It is still remembered for *de vez en cuando* (now and then) someone will sing of the exiled Moors. It never fails to bring a lump to the throat of all old *paisanos* who remember the lessons learned at their mothers' knees.

Riches . . .

Mas alla (further on) and the Spaniard burning with the fever that only gold or silver could cure, would plunge on into waterless deserts and steaming jungles. The Indians must have found a grim humor in the Spaniard's credulity. All over the Americas are graves where they lie, having perished never reaching that goal that was just a little *mas alla.*

The legends of lost mines probably emanate from the effort of the Indian to stay alive and add fuel to the white man's greed, and in this way perhaps gain a little larger dole of maize and keep the spark of life glowing a little longer. Devious and ingenious were the schemes of the red men to eke out an existence with every man's hand against them, the legitimate prey of all adventurers.

An old tale that has a touch of irony is the one of the old Indian who, not wanting his race to die out, had 20 children and in order to provide for them he wisely made a *compadre* of each well-to-do rancher in his neighborhood. They, not wanting to see their godchild suffer, would give him a sack of corn now and then, there being 20 godfathers and godmothers, he was well supplied!

The Hard Way . . .

As a kid Bill Lamas was fired by the ambition to emulate his father. He pestered the older men about catching horses or cattle for him to ride. He would climb aboard almost anything they would catch. But one day he had an experience that almost cured him of the habit.

Bill and Angelo Montana were riding from Buttonwillow to the Estrada ranch and passed through a section where a number of mares and colts were running. There was an active, fine looking, chestnut colt among them and Billie said he would like to take a setting on that chestnut.

Angelo was a quiet man. He said not a word but took down his rope and waited. The curious horses came up to the riders to sniff noses with their mounts. When the bronco was in the right position, he threw the loop without whirling and caught him by both front legs. The colt hit the end of the rope so hard he did not move after he fell. The two dismounted, took their ropes off the motionless animal and rode off in a hurry.

Having visions of the company hauling them into court and possibly sending them to jail, the two kids planned to draw their time immediately and get across the border. The boss was at Estrada. They got their checks and started back early the next morning, when passing through the field they ran onto the same colt quietly cropping grass. They had only knocked him out!

Diableros . . .

The bucking horse rider, who could ride a pitching horse until he quit, with a *peso* placed on each stirrup and held under each foot without losing the coins was not uncommon back in the old days. So old timers have assured me.

J. J. Lopez once told of a rider named Montana who could reach down to the ground and pick up a handful of dirt while the horse was bucking. Salvador Rodriguez spoke of Juan Perez who when traveling, would get fresh horses by catching a wild horse on the range and after changing saddles he would continue his journey on the wild one. *Cachequero*, at the age of 100 year was still breaking horses. And then there was Tom Williams of Lynn's Valley, a famous balance rider in his day.

Although it was not a subject to be discussed often or too openly, it was whispered that the exceptionally clever and expert *ginetes* and *amansadores* (tamers) were *diableros,* men who had made a compact with the devil for power over horses. But now and then around the campfire some wise old vaquero would venture to broach the subject of *diableros.* In order to become one, there was a preliminary discussion with his Satanic Majesty. The would-be *diablero* would find himself in a long cave at the far end of which would be a light. As he would proceed toward the light he would be menaced by two grizzly bears. These would stop and retire when almost upon him if he showed no fear. A few paces on he would go through the same experience with two lions. Further along two wolves would bar his way and finally two huge rattlers. This would prove his mettle. He then would come to the light at the end of the cave where he would find hanging on pegs such things as spurs, hackamores, riatas, bits and other articles. A little old man was custodian of this subterranean tack room and he would tell the adventurer to choose one of the implements. If he chose spurs he would become a great *ginete* and no horse would throw him. A riata would give him great skill in roping. By the selection of a bit he would be an *arrendador* or reinsman.

Having chosen his gear he would leave the cave and come out on a plain where another old man would be waiting to tell him that he was now a full-fledged *diablero*. The *sabio* (wise one) would then end his tale at this point.

The theme of the man selling his soul to the devil for a proficiency is a common one in folk tales and has given rise to a recurring form of speech, such as "he can ride like the devil," or "he is the very devil of a hunter," etc. In this country we have had some great horsemen and riders, men who could truly ride like the devil.

The *Californianos* however seem to be the only folklorists who got the Devil's aid without paying for it by subsequent servitude to Satan.

Old Avelino . . .

It was along about 1919 or 1920 when I was working on Tejon that one of the tunnels between Tehachapi and Caliente caved in. The last beef herd for that year was ready to ship and Caliente as a shipping point was out of the question so we drove from Tejon up over the mountain across Cummings valley and into the pens at Tehachapi for loading. It was quite late in the season and a cold wind blew all through the drive. The cattle too gave us a lot of trouble at the corrals. They tore down one side of the pens and some scattered on us. Three of Hill's riders came over to help us. One of them stands out in my memory. He was a little dried up gnome of a man so old that his eyebrows and lashes had long since fallen out and his beady little eyes seemed to be the only things alive in his shriveled old mask of a face. But when he rode up on a little sorrel horse I noticed that his back was still flat and he sat as erect as a general on parade. He talked to me there as we held the cattle and I learned that he was Avelino Martinez of whom I had heard much. He was then a hundred years old and still an active rider. When he dismounted to fix his saddle he didn't seem to be more than five feet tall. Afoot, Avelino suggested a bowlegged old spider. A few years later when I worked for Russell Hill we slept in the same bunk house and I got to know him quite well.

There were a lot of tales of Avelino and among them is one told to me by Adolfo Encinas. Many years back when the Tejon was still a sheep ranch, Avelino had a band on the desert side of the range. He had his camp on Alamos creek and one day some riders came to his fire. They were *ladrones* (robbers) on the run from Los Angeles. At that time the *paisanos* were in sympathy with the *ladrones* and almost always helped them and even looked upon them as revolutionists. Well the outcome of the visit was that Avelino undertook to go to the pueblo and buy them a packhorse load of supplies including ammunition and a new rifle. The *ladrones* took over the sheep but Avelino's dog would not stay with them but instead followed his master. On his arrival in town he had no trouble in buying and loading the stuff and was as far as Cahuenga pass on his way back when a posse of vigilantes halted him. They were a mixed group and included native Californians. On being questioned by them he said he was a herder on his way back to his camp on the Tejon. But the new rifle which he carried in plain sight caused them to doubt his story. So to get the truth out of him they hanged him to the nearest tree and only when he was almost dead did they cut him down. He did not change his story. Still in doubt they hanged him again but he still stuck to his story. It was the dog's presence that saved his life. One of the posse reasoned that no robber would have a dog with him and so this man must be a sheepherder. After a third hanging they gave him a document and let him go. He met more vigilantes but the document got him through. He rode hard until he came to a ranch on the desert where he got some tallow to grease the rope burns on his neck.

Avelino came from Sonora, the mother state of California, and he most likely came here while this was still part of Mexico. He spent most of his years around Tehachapi. The saying was that Avelino was so old "the pebbles had grown into cliffs" during his lifetime. He died in 1937 or 1938.

The Mexican Horsetrader . . .

It would be interesting to know what the destinies were of the numerous droves of horses moved out of California in the early days. Some of them we know were sent to Hawaii where the cowboys are called *paniolas* from the word *espanol.* California vaqueros were the first cowmen in the islands.

Explorers of the northwest found Spanish brands on horses as early as the 1830's. Cavalry mounts were obtained from a ranch in Kern County in the 70's for use in Oregon, Arizona and in Mexico when American and Mexican cavalry joinea in pursuit of the Apache Vittorio and his famous band. This brings to mind the story of the Castro mules which were sold into Mexico but at a much more recent day. They became cavalry mounts though not before a long and bitter battle on their part. I've remembered this tale as "Frenchy" told it to us twenty years ago.

Back in the times when raising horses or mules was still a profitable business, feed being considerably cheaper and much more plentiful than nowadays, one of the Castros ran a herd of mules up on the slopes of the Breckenridge mountains. Small, active, wiry, bred in one of the roughest sections of the Sierra, the animals were ideal for the eventual use they were put to and would have delighted any packer then or now. We had come from a rodeo and were recalling some of the events when "Frenchy" said: "Aw, you fellows never seen a good rodeo. You should have been with me in Juarez the time I sold the Castro mules to the Mexican army. You know those mules were bucked out in every Rodeo in the country either out of the chutes or in the wild mule races and were good bucking stock when I bought them, about three carloads in all, and all smooth mouthed. I intended to sell them for cotton mules so I shipped to El Paso, Texas. There I found a colonel of the Mexican army buying remounts for his regiment quartered across the Rio Grande. He liked the looks of the mules and after he bought them

I told him they were broke but I did not tell him for what special job they were educated. I've always figured all's fair in love or a horse trade. He invited me over to see his men on mounted drill the next morning. I wasn't too eager to accept his invitation. I didn't know what might happen when those mules turned it on. But the next morning I got up nerve enough to cross over into Juarez confident a horse trader like myself could talk himself out of any situation. The mules by that time were over in the pens across the river. On arrival I took a seat beside the colonel on the top rail of the corral. In a few minutes the show began and I've never seen a more expert and thorough job of riding and breaking on the part of the men or a more wicked display of bucking by animals grown old at the game.

"Four men marched in carrying a railroad tie which they rounded off with machetes and planted, well tamped in the middle of the corral. Then a *cabo* (corporal) entered followed by some 10 men each carrying a big horned almost cantleless saddle common to that country, a rope and a hackamore. The corporal stood out away from the men and motioned to one who stepped out and roped one of the mules. The animal, long since wise to a rope, once caught, never tightened it but led up to the snubbing post. Like an old campaigner, he reserved his strength for the coming battle by standing quietly while he was blindfolded, saddled and mounted.

When the rider pulled the blinds that stick of dynamite wrapped in mule hide exploded and went up into the air. In two or three jumps he had the rider off and the *cabo* motioned to another man who got on. He didn't stay long either but the third man stayed until the mule quit bucking, whereupon the gate was opened and the rider urged his mount out on an open field where he was left to shift for himself. But the next rider was good, and he rode his mount to a standstill. So it went, one mule after another was roped, snubbed, saddled and mounted. As the show went on some mules bucked as many as three men off but good riders were in the majority.

A good two-thirds of the men rode to a standstill on the first attempt. They had started on the third mule when the colonel offered to bet on his men against the mules. We sat there through the whole exhibition, he betting on his men's riding and I on the mules. It was in the middle of the afternoon when the battle was over and the last mule was ridden. Then each mule with a rider on his back was headed south in the cloud of dust. When we came to settle our bets I found the colonel had won most of them and it dawned on me that I had met a smarter horse trader than I was myself. He took me to dine with him that evening and in the course of our conversation I gathered that he thought like the old cowboy who said that he'd rather get bucked off by a good horse once in awhile than ride a horse that had to be prodded along all the time. And that was the end of Frenchy's story.

At one time officers in the Mexican army studied at the French cavalry school, one of the best in Europe. Since cavalrymen in general are the best practical horsemen and are the authority through testing in the field and on campaign on all questions regarding horses, it is not surprising that "Frenchy" came out second best in that particular horse trade.

Mariano's Story . . .

From before daylight that July morning we had been swallowing the dust behind the beef herd until the last one had gone through the pasture gate at the home ranch. When Mariano and I rode out of the cloud into the presence of Don Jesus a coating of mud covered our faces where the sweat and dust had caked. We had had the worst position of the drive and as a compensation the majordomo pulled us out of the crew and sent us to the pool under the gum trees to bathe and loaf for the rest of the afternoon. We lost no time in riding up there and plunging into the soft sulphur water.

Mariano was a newcomer to the ranch and since his arrival we had paired off, bunking and riding together as the men do on the ranches during rodeo time. From the first I had been struck by his devil-may-care attitude towards everything in general. He seemed to have a total disregard of hunger, discomfort, long hard days and good or bad horses, for he was always cheerful. I had often thought he must have been brought up in a stern and pitiless school for him to look on the work of a cattle ranch as something to be scoffed at.

After we had scrubbed the *mugre* (dirt) off and were clean again, we sat down on the edge of the pool to throw pebbles at the lizards which darted up and down the trees on their eternal pursuit of flies. We were sitting side by side, when Mariano turned his back to gather more pebbles. I too turned towards him and saw that his back from side to side was covered with deep scars that had been made by a rawhide whip. He saw me staring and so as we sat under the gum trees that hot afternoon, he told me the story of how he had come by the

"Why Does He Lash a Man For the Stealing He Did Himself"

marks which he would carry to his grave, and of how he had once been a Jean Valjean long ago in Sonora when the Yaquis were still gamely resisting the *Yori* (White destroyer).

Mariano's regiment had pushed into the Bacatete and the further they advanced, the slimmer had become their rations. He had been hungry for days when one morning as he came off guard duty and espied at the general's cook's place a basket on an outside stand. When the woman left the room, Mariano stole up to the hamper and lifted the cover. Wrapped in snow white napkins, was the general's breakfast. *Atole de chocolate,* eggs, tortillas, as wide as a hat and thin as a finger-nail, toasted as only a Sonora woman can toast, wheat flour tortillas, and frijolitas, fried in lard and mashed to a smooth paste.

There is an old maxim often quoted and often practiced by the men who rode the ranges in the old days to the effect that *Cien anos de perdon, a el que roba un ladron* (One hundred years of pardon for one who robs a robber). Mariano ate the general's breakfast with great *gusto.* That gentleman, finding his basket empty, accused his orderly of eating it and calling Mariano, who had lingered near the scene of his crime, ordered him to hang the culprit by the thumbs and give him fifty lashes. Mariano obeyed with a sad heart. How was he to beat this poor breakfastless man, when he had the stolen food in his own belly? He reluctantly hung the orderly up to the rafter but let him stand flatfooted instead of raising him on his tiptoes, as was usually done when this punishment was given. He laid the strokes in with a light and gentle hand until the general who was in the next room, not hearing groans, ordered him to lay on harder.

At that moment the serving woman returned and said, "How is it that he beats a man for the stealing which he must have done himself."

When the general asked him if this was true, Mariano answered that he had eaten the food. So they hung him to the rafters until only his toes touched the ground and gave him the fifty lashes. When he recovered enough to ride, he saddled a *Zebruno* (zebra-striped buckskin) mule and came to California.

Mules Vs. Men . . .

Often around the campfire some ex-soldier in speaking of his experiences while on active service would assert that the government valued and took better care of the mules than of the men. A mule cost the taxpayers a hundred or more dollars, whereas the soldier cost nothing. Sarcastic foremen on ranches often said the same thing. It was easy to get men but a good horse or mule was hard to find. A good horse cost money; a *bindle-stiff,* nothing.

Perhaps tragedy would not have struck on the ranch in the hills that border the west side of the valley, had not the foreman one day told old Kim, one of the two handy men on the place, when he swept the bunkhouse before he fed the horses that he should forget the men and take care of the horses, a rebuke which the old fellow took to heart. Or, perhaps, if the foreman's wife had not used her sharp tongue, the roustabout would not have burned to death. It all began when Kim's partner who fed and milked the cow set a match to a pile of trash near the buildings and caught the range on fire. How he managed to lead the cow out of the corral, no one ever knew. When the flames had run their course, no trace could be found of either man or beast. The brindle cow and her keeper had vanished and were never seen again.

Three vaqueros, two men and a boy, had driven the long-horned Chihuahua cattle many miles east of the road which passes Devil's Den. The boy looked back and saw the smoke billowing from what seemed to be the ranch buildings. The three men stared at the black cloud for a few moments; then, abandoning the cattle, they put their tired horses into a run back the way they had come. They had loped the horses twenty-five miles before they came to the road again, where they unsaddled and turned the animals loose. They put the equipment on the ranch truck when it arrived and rode to the fire. The weary horses drifted on to find water at the Alamo where a week later the vaqueros found them. They had encountered good grazing, and were sleek and well-rested.

At the ranch the men on the fire truck found old Kim lying in a canyon, his clothes, except for a patch which stuck to his shoulder, had been burned off. In his hand he still clutched about a foot of what remained of the bridal reins. A blue roan horse, his hair and much of his hooves burned away, suffered in mute agony until a bullet put him out of his misery. The men put old Kim on a stretcher made of two coats and two poles, but when they gave him a drink he revived enough to ask them to remove what remained of his clothes because he wouldn't need them. After a few sentences he could talk no more and died before they could carry him out of the canyon to the waiting truck.

The vaqueros reconstructed the tragedy from what Kim said and from what the foreman's wife told them. When the fire broke out the two men had stood about not knowing what to do. The foreman's wife said that if her husband were there, he would go out and fight that fire and not hang around the house until everything burned. So Kim caught the roan horse, that was left in the corral, because he was mean and no one wanted to ride him. He saddled the animal and hung a number of sacks on the saddle with which to beat out the fire and mounted. The horse that bucked at every chance, strangely enough this time paid no attention to the burlaps that brushed his legs but walked off like a lamb.

Had Kim ridden down the canyon above the fire over the burned ground he would have been safe. But he rode into the path of the flames up a narrow gully. He dismounted and took a sack to beat out the fire. In a few minutes the flames reached the horse which refused to lead into the burned ground. The sacks hanging from the saddle caught fire and perhaps remembering the foreman's words, the old man was trying to unsaddle and save the horse when the smoke overcame him.

The three vaqueros in going over the ground found the buttons from his pants and shirt in one neat pile where they had fallen as the flame consumed the clothing. Thus, old Kim met his end because of a man's sarcasm and a woman's sharp tongue.

Curanderos . . .

In the old days only the *curandero* (healer) practiced the art of curing disease. Like the noted *ginetes* and *arrendadores*, they had arrived at their art through the study of black magic. Being a jealous lot, these sorcerers were often at odds with each other and would pit their respective powers against one another to determine who was the stronger magician.

One tale often told was of the healer who was so rich, fat, and lazy he never would stir from his fireside. Don Panfilo was ailing, and having heard of a *curandero* who had much skill, he sent a vaquero to guide him to his bedside. The vaquero led a horse to mount the healer whose burro was deemed too slow for so urgent a journey. After much coaxing and many promises on the part of the guide the reluctant quack mounted the horse and set out with the vaquero, though not without grumbling.

After traveling a few miles he found the animal a hard trotter. The horse's gait caused a sharp pain in the fat man's side. To relieve it he would stop every little while, dismount and sit under a tree. After many such rests the vaquero began to fret at the slowness of their progress.

They came to a spot where the bleached skeleton of a bull lay under a tree. Here the curandero dismounted to rest again. The vaquero, who had lost patience and hope of ever getting him to mount again, decided on stern measures. When the fat man started on his second cigarette the vaquero, who was a magician in his own right, said, "*Curandero,* get ready, you are going to fight a bull". The other answered, "I see no bull, only this heap of bones." "Well, here he comes," his guide warned as the skeleton rose and charged the sitting man who jumped up and dodged.

The skeleton charged again and again, tearing his adversary's cape to shreds. Finally the dodging fat man cried, "Enough! Enough!" and the skeleton sank down again into a heap of bones. When the *curandero* had regained his breath he mounted his horse, and turning to the vaquero said, "Now it is my turn to make you jump."

He waved his hand and immediately a flame arose which moved in a circle around the vaquero. The fat man rode away and called back to the man in the circle of fire, "Jump out of that if you can!"

Loosing a Cook . . .

Though shunned by all men the rattlesnake served a second good purpose besides eating rodents. Sometimes a cook became so bad that the men were sick of the food, everyone's stomach was upset and all went hungry rather than eat. If, despite the grumbling of the men the cook stayed on, one of the riders would catch a rattlesnake, put it in the cook's bed then roll it up. When the cook discovered the snake he would take the deadly hint and betake himself off the premises.

Canny Old Miller . . .

It was back in the days when the ranchers of Kern County suffered from a plague of horse theives who robbed rich and poor alike. No man's horse was safe. The strapping sons of Don Tomas Castro had finished breaking the first band of colts bred from mustang mares and a Hambletonian sire. They were good, serviceable animals and with justifiable pride Don Tomas had offered them to Henry Miller.

In expectation of a good price and praise for their work the entire Castro family was on hand to greet the cattle king who on arrival went at once to the corral. Leaning on the bars he looked the horses over carefully and turning to one of the boys he pointed out a bay saying, "Catch me that one." The boy entered the corral eager to show how gentle the colt was. It stood quietly while he put a rope around its neck and lead it up to Miller who frowned and asked, "Are they all that gentle?" On being assured that they were he said, "I don't want horses that gentle. They can be caught too easily and horse theives would soon steal all of them. They are sheep herders' horses."

The crestfallen Castros watched the old man drive away in a cloud of dust, their disappointment not unmingled with a great respect for his canny reasoning.

Old Miller Again . . .

The trail is long, dusty, and slow from Buttonwillow to the Mendota Slough where Antonio Maria Lopez, the boss, had stopped to rest his herd and the vaqueros who drove it. They awaited Henry Miller, who was to meet them there. The cattle king arrived, hurrying his buckboard team into camp. This was his habit. His horses had no sooner come to a halt than he threw the reins on the ground and jumping out of the rig went immediately to inspect the cattle, in that anxious manner he assumed when a question of loss or gain was involved. After Miller finished with the cattle he ordered the wrangler to drive up the saddle horses where he could look them over. Of all the band only one horse had developed a sore back. Upon inquiring, the cattle king learned it belonged in the "string of the boss himself, Antonio Maria Lopez.

The men's saddles were all lying in a row by the corral gate, taking the wrangler with him to point out Lopez' saddle *Old Henry* picked it up and carrying it to the fire threw it into the flames.

Turning to the men who had gathered to watch the performance, he signaled out *Canducci* (Emeliano) Castro who was then a young boy. Miller ordered him to give his saddle to the *mayordomo* and to ride on the wagon with the cook for the rest of the drive. With that the blunt old man climbed into his buckboard and hurried away on his endless tours of inspection.

Some days later the vaqueros arrived with the trail herd at the Canal Farm, where Antonio Maria Lopez found awaiting him a brand new, full flower stamped Visalia saddle, a gift from Henry Miller. On the tapaderos was carved HH the cattle king's favorite brand.

Corned Willie After Hay . . .

It is when rodeos are in progress on the ranches that conditions of distance and terrain cause the buckeroos to go without food for the longest periods. From the time they leave camp at daylight it is a long hungry day until they return, seldom if ever, before dark. They often go thirsty, as well water is scarce on most Kern ranges. The horses suffer most and it is hardship endured together which develops comradeship between a man and the horse he rides. It is when both man and animal are weary and the rider wonders if there is one more spurt left in his jaded mount, which responds on leaden legs that there is born the respect a vaquero has for a sturdy animal.

One may keep a pet horse to dote for its lifetime but it will be only in the test of hard work and emergency on the range where a horse is found to be truly noble. It is an old and well established rule that a vaquero feed and water the horse he rides before he seeks food and water for himself. This is the reason a rider seldom carries a canteen or a lunch on his saddle. No man with sporting blood in his veins will let even the worst mule go neglected. In good truth the great majority abide by this rule and those who do not are rare exceptions.

When some green kid in his ignorance ventures to provide for himself some old timer will ask him what he is carrying for the horse to eat or drink as the case may be and will then inform the youngster that if the horse which is carrying him can stand doing without he who is riding should be able to do the same.

I learned this years ago when I was a kid on a trip I made with Rafael Marvis of Poso Flats. It was said around Greenhorn mountain country that the riata which Rafael carried had eyes and when he threw its full length of 60 feet the loop would find its mark and settle around the neck or feet of the animal no matter how or from what position he threw it. One old timer said Marvis could have taken a work mule out of a plow harness and used it to rope on with success. He was in late middle age and quite fleshy when I first met him and to look at him one would never dream he would be the Spartan he proved to be in following the established rule.

A horse had fallen with Oscar Clendenen who had suffered a broken leg in the accident. Oscar was still in a cast when branding time came around that spring. I happened along and was hired to go with Marvis, his partner to gather the stock then in the hills along Poso creek. A pack horse would have been impractical for so short a journey. On the morning we left Poso Flats with no provisions whatever, Rafael's sister called to us and brought out a package wrapped in a sack. It was a can of Corned Beef then sold as war surplus. It was the corned Willie of World War One, and the subject of much reminiscing and no little griping by veterans of that war. Rafael tied it to his saddle, with some reluctance, it seemed to me.

The shack we came to that evening had a stove in it and a mattress rolled up in a corner, but not food of any kind. There was a corral in back and a shed but no hay in it. We put the horses in the pen to pass a hungry night. When we had made a fire in the stove I asked Rafael if I should get the can of corned beef. He said no, that the horses hadn't eaten and that when we found food for them it would be time for us to eat. "And besides," he said, "we may be out four or five days and better save it until we are really hungry."

I went to sleep on the mattress on the floor hoping we would find hay for the horses soon. From then on until we ate the corned beef I was on the anxious seat. I worried about it coming untied from the saddle and being lost. Whenever we separated and met again I looked to see if it was still tied in place. The next day we found some of the cattle and drove them to a place where we found hay and on the shelf in the shanty some flour, enough to make one cake apiece. This we ate, saving the corned willie, and consoled by the fact that the horses at least were well fed.

The third day we cleaned up the cattle in that section of the hills and that night stopped at another deserted house where Salvador Rodrigues and a young Castro boy were awaiting us. There were some beans there which Sal had cooked and with the corned willie we had a feast, there being hay for our horses. Salvador and the Castro boy helped us. On starting down a long hill the pinto Castro was riding lit into bucking and when I headed my horse toward it to pick it up, he yelled to let her go. A salty boy was that Castro kid.

We made it back to Poso Flats on the fourth day. This happened almost thirty years ago and I've forgotten a good many things that happened since but the can of corned Willie and the lesson in sportsman ship I learned from the good vaquero Rafael Marvis will always stay fresh in my memory.

Spanish Gringos . . .

Once there was a time in California when Spanish was the dominant language and the traces of the linguistic pattern are still dimly apparent and were more so not so many years back.

Before the paisano had modified his ways to those of the *Americanos* Spanish was the common tongue. The *hijos del pais* were not compelled to use English, which to them is a harsh *gringo* tongue that never sounds like it is written and must be spoken with so much care in order to avoid ridicule. There were here then several Americans who were not looked upon as strangers by the natives, but as kindred souls, for they spoke fluent Spanish. Some of them by virtue of personality had attained the title of *Don*.

Tom Briggs was known as Don Tomas, Bob Glenn, constable at Tehachapi for years and probably the best informed man on early day events told many tales of those times. He not only spoke but could read and write the most flowery Castillian. Canducci Castro tells me that his boyhood hero, Jim Crocker, spoke better Spanish than Don Thomas Castro. George Coffee, an early day pioneer left a tradition of horsemanship to his descendants, June and Nan Uhalt, who are very active in present day horse shows. He spoke Spanish well. Walter George Harrison of Miller and Lux and superintendent for the elder Hearst was much sought as an interpreter and his genial son Homer still retains a smattering of the tongue and speaks intelligently of *Manganas, piales* and *lazos*. Most of the McKenzies of Tejon canyon district and the Winters boys learned the language from the vaqueros on the ranch. Archie, a member of the family speaks with all the little peculiarities common to the idiom spoken there and is accepted as a member of the Tejon rancheria. William Clark of Tehachapi and all the Tibbetts except Bert and Ed were bi-lingual. Then too there was Tom Watson who was fluent in Spanish. With Perfecto Cuen he joined Buffalo Bill's wild west show.

While some from respect were addessed as *Don* I never heard but one American called the intimate *Vale*. Russell Hill called Rosalio by the *paisanos* one day, after a trying experience, gave vent to his pent-up feelings in such robust Spanish that all the vaqueros roared with delight and Ramon gave him the accolade *Vale*.

He said, "You are now a full-fledged vaquero as you express yourself so forcibly." A first class reinsman and spade bit man, Russell always rode a fine Morgan horse and his stock was considered tops in that breed. The drought of the last few years caused Mr. Hill to move his blooded horses and cattle to Madera. His going is a distinct loss to Kern county.

Ed Conley, a cattle buyer, was another who understood the *hijos del pais* and was very adept at the language. One day when he was parting cattle at White Wolf Ranch, Mr. Lopez, the superintendent, kept saying, *Parate, Parate* (stop, stop) to the milling animals which kept moving and were hard to hold. Conley asked the boss if he thought the cattle understood him. Don Jesus puffed out his cheeks, bristled his mustache until the ends stood out and finding Spanish inadequate said, "Meestaire Conlee, theese cattle know my voice. I raise them." The idea of those steers, some six and seven years old, a cross between an antelope and a wild cat, knowing the old man's voice brought covert grins to the buckaroos. It was in my associations with Don Jesus that I began to understand why the Spanish people have never changed the original text of Don Quixote and why after 300 years Cervantes is still read with delight. The battle worn old soldier knew his people.

Clandestine Bronc Riders . . .

As one of the foremost early-day cattlemen in California Henry Miller's success in his enormous operations was due in a great measure to his ability to pick good foremen. Cattle kings of his stature required henchmen like the old iron man, Rafael Cuen, who could not fail in his fidelity to the charges under his orders. He could get more work out of a crew, good or bad, than any early-day cattle boss. His one aim in life was to move cattle. Driving his men was only incidental to attaining his ends.

While some of the best buckeroos in the country worked for him at one time or another, he often found himself with a poor, inexperienced lot, but regardless of circumstances or conditions in his forty four years of service he was one of Miller's most competent bosses. Though his legs had long since become curved to a horse's barrel, by the time I first saw him his back was the straightest and his shoulders the squarest of any of the fine old-time horsemen. His mustache was always trimmed and he always found time to keep clean shaven. Another vanity was his boots, handmade by Carl A. Adams in Miller's Bakersfield shop now long out of business.

The following story shows why Cuen remained a boss for nearly half a century. It was in the early twenties and Johnnie Puget, Bill and

Fred Lamas and Bob Huntington were working on the Miller wagon. They were all kids then, and burning with ambition to be great bronc riders. One day when Cuen had left early to attend his many duties on the great ranch, the quartet ran some steers into a corral and took turns bucking them out. They got a lot of practice and had a lot of fun. Cuen got back after dark when all were asleep. The next morning they were eating breakfast, when the boss, to their surprise, told them bluntly to eat heartily because the work that day would not be a picnic like the one they had had the day before riding the fat off the company steers. Later in the day the four *companeros* went into a huddle to try to solve the mystery of how "Old Jake", as Cuen was called, had found out about the steer riding. None of the four had spoken to him since he had returned to camp. It remained a puzzle to them until they again rode some more steers in "Old Jake's" absence.

That night all but Bill were asleep when Cuen came into camp. Bill, wide awake, watched while he got out his bedroll and spread it next to Johnnie, who was snoring away. He sat down on it and rolled a cigarette as if waiting for something. Pretty soon Johnnie started to mumble and the boss got a full list of the event of the day from the kid who talked in his sleep. The next day they all knew who Jake's informant had been and whether they devised a gag for Johnnie to wear while he slept or whether they quit trying to fool the cunning old boss, I never learned. But in later years all four of them acquired some fame as bronc riders.

A Homing Horse . . .

In the stories of years ago it is brave horses that are the main characters in the memories of the old-time riders whose lives were just a series of different horses.

The horses of old, they say, were of perfect shape and size for the work and conditions found in those days, carrying blood of Barbary sires transmitted through Andalusian dams, descended from the horses brought over by the conquistadores. From the very beginning of their entry into the new world, they were never other than heroic from the numerous charges they made at Tenochtitlan against the Aztecs till they became a memory to the old riders.

The bucking, fighting mustang was tireless, enduring and almost unkillable. It is an historical fact that horses needed no shoes here until the advent of the American with his horses.

The first horse shoer in Los Angeles was an American. A Spaniard from the horse breeding section of Spain once told me that the tears shed by Boabdil, the last Moorish King of Granada, were not shed for the palaces he had left behind, but for the horses he had lost.

Now it may be an odd thing, but box stalls are not often found for range horses on cattle ranches and these horses are seldom groomed. It is a rare sight to see a vaquero currying a mount and knowing this I once asked Adolfo if he groomed his string in the old days.

He told me, "No," but said he could tell me a tale of a horse that not only was groomed in the old days, but was stabled every night in the box stall with fresh bedding. How Don Jesus (J. J. Lopez of the Tejon ranch) discovered the peculiar aptitude of the brown horse, or how he ever took the risk of letting the animal prove his homing instinct, I don't know, for Don Jesus loved his horses.

Long before I was old enough to ride at Tejon, supplies were hauled in a wagon and trailer pulled by eight mules. The trip was made twice a month but mail was delivered oftener and this is how it was done.

There was at headquarters a brown horse called *El Cartero* (mail carrier) and whenever a buckeroo had earned too much money and the bright lights of Bakersfield were beckoning the plutocrat would ask Don Jesus for a few days in town. He would be told to turn his string out with the *caballada* (saddle stock) and his check was made out. He was told to saddle Cartero early in the morning and start to town on his way leaving the gates open. He was instructed further upon his arrival in Bakersfield to go to Newsom's livery stables, see that the horse was watered and fed grain, go to the postoffice, pick up the mail, return to the stable and after the horse had been fed to bridle him, tie the bridle reins securely to the horn and likewise the mail bag. Then the horse was led to the back of the stable and turned loose with a slap on the rump.

The riderless horse would start off at a gallop and never let up until clear of the town when he would walk for a breather, then gallop again. He never stopped on the trip even if he met or passed other horses, but kept a straight course until his mission was discharged at the home ranch. Now and then, someone would try to catch him as a test, but he was *muy mecha* (hard to catch) and if cornered would knock his would-be-captors aside and go on. This is how Cartero earned a box stall with fresh bedding to sleep on, so said Adolfo concluding his story of the *mail carrier*.

Now as to this horse, he was a *barileno,* a fine strain of mustang found on the Tejon at that time. The name, I think, is derived from barilitos, Barrel Springs in Antelope Valley where a herd of wild horses ran. The colors were usually buckskin, a dark palomino and mouse color.

There are other interesting stories of these mustangs. Don Antonio Leivas, one day, when we were *campiando* (riding the range) told me of the last wild stallion, a palamino, of that herd. He was run for days. When at last he was caught in a loop set to trap him, he killed himself fighting the rope.

He also told me that early one morning he saddled a *barileno* at Cummings ranch in the valley of the same name. He mounted and got past the house when the mustang went to bucking and kept it up until he was tired. Then he balked until he was rested and bucked again. He kept this up and it was late morning before Don Antonio finally got him into a run. That night he was in Los Angeles a little after dark!

El Rodeo . . .

To *parar el rodeo* (hold the roundup) means to stop and hold the cattle that are driven in from the surrounding territory. The *puestos* (posted riders) hold and *mill* them until they come to a stand and quiet down. Then the foremen (usually) ride in and part out the best steers for market. The cattle are very restless during the start of this operation and break away from the herd, causing no little running on the part of the *puestos.* In all vaquero crews there are to be found *matreros* (shirkers) who, when an animal makes a break for liberty, always wait to see if some one else will start after it. Usually old men, whose bones have become too stiff, find no joy in running their horses and contrive to place themselves between two youngsters who are eager to run their mounts. *Viejos matreros* (old shirkers) the youngsters called them, but under their breath, of course, because old age was respected. Even when justice was on the side of youth, a *matrero* (shirker) was preferable to a *grosero* (insolent one). Disrespect was abhorred by every one and the saying, *"Pues es viejo"* (Well, he is old) excuses everything.

Adolfo, Catarino and I had been reminiscing about the old days in general and about Don Jesus Lopez in particular. Adolfo had been one of the old Don's *hijitos* (little sons) and of course remembered many incidents in his association with the old man. He and I were surprised when Catarino related a little story concerning the Tejon, because neither of us knew he had ridden on that ranch, having put in fifty four years at Bellevue. Most of that time he was a foreman.

Well, to go on with his story; the vaqueros had been gathering beef and had a sizable herd ready for the buyer when Catarino and the rest of the crew were ordered to night herd the cattle. In those days the management of the Tejon scorned such effete accommodations as corrals and fences. To their way of thinking, vaqueros were made of iron. When darkness set in the men started riding around the herd and after some time Catarino noticed that he met only one rider, Fernando Cordero, in his circles around the *bedding ground,* but being young, he cheerfully went about his task and would have soon forgotten it, but the next day the entire crew was riding in a body when they came to a gate. Mr. Lopez, who was riding in the lead urged his horse forward, stopping in the middle of the gate and turned him around to face the men and held up his hand for them to stop. The surprised crew halted. Don Jesus signaled out Fernando Cordero and Catarino and motioned them to ride through and on. Thus having separated the sheep from the goats, he bristled his mustache and shaking his finger at the vaqueros, he proceeded to give them a lecture on slothfulness and the danger of sleeping where nervous cattle were being held, called them *flojos* (lazy ones) *matreros* (shirkers) that laid at their ease while others worked. Catarino and Fernando never learned how the old man knew the rest of the crew *matreriaron* (had shirked).

The Hard Way . . .

In the old days boys learned to be vaqueros the hard way. When ever a beginner was either too slow or was hesitant about the work around the rodeo ground, some old vaquero would ride up behind him and start lashing him with a reata (*reatazos*). The old man managed to give him two or three good cuts before the boy could spur his horse away from the punishment; or, if he clutched the saddle horn when riding a bucking horse, a few blows with the butt end of a quirt would teach him to ride without that aid.

In those days, too, a grown person felt himself duty bound to punish any youngster unlucky enough to be caught in an act of mischief or deviltry. After giving the boy a licking, the older one would order the kid to report to his father, whereupon the father would administer another *reatera* for allowing other men to see him in his *sinverguenzada* (shamelessness).

The Woman and the Steer . . .

While a man on horseback may ride among cattle without causing them concern and may on occasion quiet their restlessness, a man by dismounting can cause a stampede and a person afoot will be attacked. Most of the trouble with cattle is caused by someone walking near them.

The wild longhorn cattle had come from the *bosques* of Chihuahua in old Mexico and were bred on the ranches of Don Luis Terrazas. For many years after the railroads came in, a steady stream of these cattle moved into California from the vast *haciendas* below the border. Miller and Lux bought many carloads and Catarino Reese was in the crew when the vaqueros at Buttonwillow unloaded a trainload and started north with them toward the Santa Rita.

The wilder the cattle the easier they are to drive, and this herd traveled fast, too fast in fact, for Mr. Reese says that quite often the vaqueros had to gallop to keep their positions in the drive. Every day the steers would string out in a long line and travel at a steady pace and the vaqueros had no trouble until they came to the long lane at Goshen.

In driving stock through a lane the riders on the point close in behind a small bunch of the leaders and drives them ahead to keep them from bunching. Two other men gallop ahead to stand at intersections or close gates that are open along the line of march. The rest of the cattle keep moving as long as the leaders do and seldom turn out of the lane once the leaders have passed an open gate or intersection. As Catarino said, the cattle were traveling too fast, and the crew was *short handed.* He was in the rear with Pedro Cuen the wrangler who was driving the horses, when things reached such a pass he told the *rango* to leave him and ride up ahead, the horses would follow. Catarino took up the rear riding in the cloud of dust behind the drags.

He was passing a grain field when he dropped back to get a breath of clean air and saw five steers in a barley field along the lane. Near by was a homesteader's shack. The San Joaquin farmer was called a *sandlapper* in those days and often had a hard time keeping body and soul together; a far cry from the *shirt sleeve millionaire* this man of the soil has become in late years. Catarino turned his horse and rode back to a gate that opened into a meager stand of barley, intending to drive the steers out, when he saw a woman with a baby in her arms come out of the shanty and walk hurriedly toward the steers apparently intent on driving them out. Seeing what her intentions were, he put spurs to his horse, a sorrel colt, and rode toward her yelling for her to go back, but the woman paid not attention or did not understand. She must be one of the many diverse races that people the great valley, the vaquero thought, and tried Spanish, but she ignored that too. He knew a smattering of Italian and Portuguese which are often met with in the San Joaquin and come easy to those who speak a Latin tongue. After a trial, these too proved useless. The woman walked on through the stunted grain toward certain death. The vaquero put the sorrel colt between her and the steers and by driving at them, turned them toward the gate. Four went through but the fifth, a brindle devil, turned away and trotted up the fence and the woman ran toward it to head it off. The steer stopped and began to switch its tail and pointing its ears forward and the vaquero knew it would charge any moment.

It was one of those life or death emergencies a vaquero meets in his daily life. If he stays alive he developes a faculty for quick thinking and rapid decisions. Catarino weighed all the chances; the colt he was riding had never been roped on, and would buck at times; if he roped the steer the horse might buck and the steer could kill the woman and baby. She had come to a stop at long last, realizing her peril. As the brindle charged, Catarino spurred his horse into a run and overtaking it, drove the sorrel into the longhorn's shoulder, toppling it over. Vaqueros call this *testeriar.* Before the animal could right itself, he gathered his horse and drove it into the longhorn again and again till he pushed it through the fence, across the lane and through the next fence where the steer fell and lay *sulling.* Catarino sat for a long time on his panting horse, shaking like a leaf from head to foot. Later, when the nervous spell had worn off, he got the steer up, the fight being knocked out of it. The brindle, smelling the tracks of those in the herd, trotted up the fence after the others.

Catarino dismounted to fix his saddle and looking back at the miserable little stand of barley, wondered if the woman who had run such a risk to protect it was really as fearless as she had seemed, or whether she was ignorant of the danger she had been in. For a long time afterwards he would wake at night and see the Mexican steer switching its tail and charging the cringing woman.

He had parted company with Pedro Cuen early that morning and it was six hours later when he rode into the field, driving the last stragglers, where Rafael Cuen and the vaqueros had turned the cattle and were holding them waiting for him to arrive.

The End

Leonardo Chavaria

Catarino Reese (Ruiz)

Emiliano "*Canducci*" Castro

Orville "*Jack*" Glenn

Pedro Yorba

Tiburcio Vasquez

Juan Valdez (*Muy Vaquero*)

Jose Jesus Lopez (*Don Quijoto Renacido*)

Pomuceno Cordero *(Indian)* Eduardo Valdez, Sutah Coway *(Indian)*

Clarence Clark, Ed. Valdez

Cholo Valenzuela

Young Vaqueros

Lassoing the Lord of the Continent

Made in the USA
Las Vegas, NV
10 November 2026

57843107R00083